Challenges of the Firefighter Marriage

Challenges of the Firefighter MARRIAGE

ANNE GAGLIANO

MIKE GAGLIANO

> **Disclaimer**
>
> Information contained in this work has been obtained from sources believed to be reliable. However, neither Fire Engineering Books & Videos nor its author(s) warrant or represent the accuracy or completeness of any information published herein and neither Fire Engineering Books & Videos nor its author(s) shall be responsible for any errors, omissions, or claims for damages, including exemplary damages, arising out of use, inability to use, or with regard to the accuracy or sufficiency of the information contained in this publication. Additionally, the views expressed in this book are purely those of the author and are based on the author's personal experience balancing a career as a firefighter and marriage.

Copyright © 2018 by
Fire Engineering Books & Videos
1421 South Sheridan Road
Tulsa, Oklahoma 74112-6600 USA

800.752.9764
+1.918.831.9421
www.FireEngineeringBooks.com

Senior Vice President: Eric Schlett
Operations Manager: Holly Fournier
Sales Manager: Josh Neal
Managing Editor: Mark Haugh
Production Editor: Tony Quinn
Illustrations by Paul Combs
Cover photos courtesy of John Odegard and Tanya Miller

Library of Congress Cataloging-in-Publication Data

Names: Gagliano, Anne, 1965- author. | Gagliano, Mike, author.
Title: Challenges of the firefighter marriage / Anne Gagliano, Mike Gagliano.
Description: Tulsa, Oklahoma : Fire Engineering Books and Videos, [2018] |
Includes bibliographical
  references.
Identifiers: LCCN 2018008301 | ISBN 9781593704469
Subjects: LCSH: Fire fighters--Family relationships. | Fire fighters'
  spouses. | Marriage--Psychological aspects.
Classification: LCC HD8039.F5 G34 2018 | DDC 306.81088/36337--dc23

All rights reserved. No part of this book may be reproduced,
stored in a retrieval system, or transcribed in any form or by any means,
electronic or mechanical, including photocopying and recording,
without the prior written permission of the publisher.

Printed in the United States of America

3 4 5   22 21 20 19

To my husband Mike, my muse, my beloved, my best friend. Thank you for making my dreams come true. This book is a love story written for you.

To my sons Michael and Rick, daughter-in-law Samantha, and granddaughter Quinn; you are my joy and comfort, delight and sunshine. I am so very proud to be your Mama (and "Grand" Mama.)

To my Mom and Dad, thank you for instilling in me from a very young age a deep and lasting love of books and the written word. Wish you were here to see this.

To my Auntie Lee: you too gave me a love of books when you sent me Little House in the Big Woods. *Then you took it one step further and taught me to love God's word. I'll always consider you to be my "Spiritual Mother."*

To my extended family, Mark and Jennie Holland; Mickey and Karen Gagliano; Tristin, Jeff, and Aurora McMahon; and Josh Ferris; I'm grateful for each and every one of you and for the love and support you give.

To my Lord and Savior, Jesus Christ, the author and creator of marriage and of the noble self-sacrifice upon which the fire service is based. "Greater love hath no man than this, that a man lay down his life for his friends." John 15:13, KJV

CONTENTS

Foreword by John Mittendorf...ix

Foreword by Diane Rothschild xiii

Acknowledgments xv

Introduction xvii

I THREE THINGS THAT MAKE FIREFIGHTING DIFFERENT FROM OTHER PROFESSIONS

1 Exposure to Danger..3

2 Exposure to Trauma 15

3 The 24-Hour Shift...29

II FIVE ESSENTIAL CONVERSATIONS FOR THE FIREFIGHTER MARRIAGE

4 Essential Conversation #1: Re-entry Time......................... 45

5 Essential Conversation #2: Harshness and Gallows Humor........ 57

6 Essential Conversation #3: Handling the Tough Runs 67

7 Essential Conversation #4: Dealing with the Fix-It Mentality...... 79

8 Essential Conversation #5: Keep Your First Family First........... 93

III HOW TO LIVE WITH A FIREFIGHTER

9 The Firefighter and Time Off....................................... 105

10 Exercise.. 115

11 How to Sleep with a Firefighter 129

12 When a Firefighter Spouse Sleeps Alone 141

13 The Healing Power of Touch 153

IV SEX: A VITAL COMPONENT OF MARRIAGE

14 Sex and Marriage 167

15 Pornography vs. Marriage... 181

V CHILDREN

16 Kids Are the Icing, Not the Cake 193

17 Thoughts on Raising Solid Children............................. 205

18 Fathers and Daughters... 225

VI MONEY
    19 For Richer or Poorer 239

VII THE TWO TOUGHEST ASPECTS OF FIREFIGHTING
    20 Firefighters and Cancer 255
    21 Firefighting Is a Deadly Calling 269

VIII ENDING ON A POSITIVE NOTE
    22 The Power of Encouragement 285
    23 Divorce Is Not the End 299
    24 Don't Stop Believin' 309
        Conclusion 319

FOREWORD

By John Mittendorf

As most firefighters are well aware, the fire service comprises an immense array of varied pieces that come together as an organization that is well equipped to alleviate a wide variety of emergency and nonemergency incidents. This proficiency requires extensive training resources that are summarized and easily accessed via the Fire Engineering Training Network, Fire Engineering books and videos, and other similar resources by numerous authors that detail the many comprehensive responsibilities of typical professional and volunteer firefighters, regardless of rank.

However, with all the abundant training resources available to members of the fire service, a quick review of these assets indicates that there is one topic that is curiously missing: marriage challenges in the fire service. It would be easy to assume that its absence from training resources means that most firefighters are assumed to have their act together in their marital lives and homes. However, current statistics paint a completely different picture: as divorce rates among male firefighters can be higher than the general population, divorce rates for female firefighters are reported to be significantly higher than the general population, and marriage counselors indicate that marriage and family difficulties are noteworthy reasons for firefighters and their families to seek counseling.

Although marriage is typically assumed to be a normal aspect of human life, marriage within the fire service is often challenged by numerous factors that are not present in marriages within the general population:

- The fire service is similar to the military in that personnel eat, sleep, and die together on a 24-hour basis.
- Firefighting is often listed as one of the most dangerous occupations. When large-scale tragedies (such as 9/11) happen, they reinforce this perception, particularly to the families of firefighters.
- Numerous health hazards, such as burns, smoke inhalation, cancer, and physical injuries, are common.
- Long hours at the fire station and the resultant lack of sleep, particularly when working overtime shifts, can mean absences of up to 72 hours. Those firefighters who are busy with EMS calls, especially after midnight, will typically make up lost sleep at home, and firefighters engaging wildland fires can be gone for weeks and longer.

- Firefighters are usually absent from their families 33% of each year, including major holidays.
- It is common for firefighters to neglect family obligations because they are participating in fire service events or working outside the fire service between normal firehouse shifts.
- Volunteers work normal jobs and then volunteer for their local fire department on their off time—days, nights, and/or weekends.

Thankfully, Anne and Mike Gagliano have authored this timely—and much needed—book, which should be on every firefighter and marriage partner's list of essential study materials, both as an initial read and as recurrent study material to enhance basic marriage foundations. I am delighted to have been asked to share my thoughts in this foreword, as I have personally known Mike and Anne for many years and feel they exemplify the ideas and recommendations they outline here. Having had the privilege to look at the book, the Gaglianos have succinctly addressed the key issues that are the basis of an unspoken need in the fire service. Although there are 24 chapters in the book, each chapter is an easy read and includes such vital, fundamental topics as the 24-hour shift, the fix-it mentality, keeping family first, financial stressors, sex, time off, and many others that will surely cause reflection and careful thought in relation to past events as well as current conditions at work and home.

After reflecting on the book and thinking back on my 30 years in the fire service and also applying its contents to my 45-plus years of happy marriage, four words kept creeping into my mind that I believe every firefighter, whether paid, volunteer, seasonal, male, or female, should honestly and thoughtfully consider: priority, thought, action, and God. Although I cannot speak for Mike and Anne, I believe they would totally agree that these four words are a basic minimum for serious reflection, thought, and consideration for any needed changes.

Priority

Interestingly, *Webster's New World Dictionary* defines priority as: "to arrange in order of importance." Where do you place your priorities? Obviously, work needs to take priority over some things, but as Anne points out in her introduction, "Marriage—There is nothing more rewarding, fulfilling, and beautiful than a happy one. It is the most intimate relationship available to humankind. It is the ultimate source of strength, encouragement, and love, and is meant to last a lifetime. Without love and family, there would be no civilization worth saving." Consider *Webster*'s definition and Anne's comments on priority; how and where do you place your priorities? This book will definitely help you evaluate the order of yours.

Foreword

Thought

Turning again to *Webster's*, thought is defined as "the power of reasoning; attention; consideration." Yes, although this book is an easy read, it can also be thought-provoking if the reader takes the time and gives proper attention to considering its contents in relation to their current circumstances, both at work and at home.

Action

Even without *Webster's*, I can tell you that the easy definition of an action is something that is done. From a practical and relevant perspective, this means that the reader needs to carefully evaluate past and present priorities, thoughtfully consider any necessary changes or modifications, and then ensure that the proper changes or modifications are made. This book will help in that endeavor.

God

Every marriage, whether secular or religious, is based on a foundation that provides a base and some type of support for the relationship. From a biblical perspective, the marital relationship was established by God as the first human institution as detailed in Genesis 2:24. In this verse, the phrase "joined to his wife" represents the inauguration of a new and primary responsibility. From a biblical viewpoint, that primary responsibility is your partner, and this responsibility carries the sense of a permanent or indissoluble union so that divorce was not considered. From the viewpoint of this book, it is crucial to keep a marriage healthy in concert with the demands of the fire service. From a personal viewpoint, to begin the marriage between my wife and I over 45 years ago, I quoted the following Bible verse, which we still honor:

> Choose this day whom you will serve. But as for me and my house, we will serve the Lord
>
> *—Joshua 24:15*

I firmly believe that the contents of this book will have an impact on many firefighters who are looking for just such a resource, as it provides a substantial foundation for strength, hope, and any necessary changes. Additionally, a careful reading of this book combined with an honest evaluation of current relationships and circumstances will dramatically improve some relationships that are currently considered average or just workable.

John Mittendorf

Battalion Chief (ret.)
Los Angeles Fire Department

FOREWORD

By Diane Rothschild

In 2010, I was asked by my bosses to come up with an idea for a product that was not hardcore training, which had been the sole mission and focus of *Fire Engineering* for the past 140 years. Thus, Fire Life was born. Fire Life is a web page geared toward the firefighter family, with articles and videos on extracurricular topics like relationships, hunting and fishing, fitness, and cooking. It was a stand-alone website for many years and is now part of FireEngineering.com.

On Fire Life, Anne Gagliano found a home for the relationship advice she had been sharing formally in lectures with her husband Mike around the country and informally with her Seattle fire service family and friends. Her article "What Every Firefighter's Spouse Should Know" immediately placed among the top 25 most-read articles on our website and continues to do so today. As of this writing, her column "HERO (to Everybody Except His Wife)" was the number-one read article for August, 2017, with 46,598 page views.

Anne's popularity is easy to understand. She is not afraid to be frank and direct about the elements of a marriage that are often difficult to discuss with others: sex, fidelity, fighting, communication—the list goes on and on. And (maybe sometimes to her husband's dismay) she does it by showing how her marriage has gone through the same trials and tribulations as other marriages.

This book, based on the wildly popular and often viral articles, will appeal to you whether you are dating a firefighter or married to a firefighter; whether you are a "probie" or a veteran of marriage; whether you are experiencing "wedded bliss" or going through a rough patch because of the stressful, demanding work of firefighting and the strain it can place on even the best of marriages. It is practical relationship advice based on years of experience. As you read it, you will find yourself hitting your forehead and saying, "They are describing us! That's us exactly!"

You don't even have to be a firefighter's spouse to relate to the advice. One of the most popular articles, "Re-entry Time" (posted October 2010), comes to mind: "Re-entry time is basically an uninterrupted period of time (about an hour or so, maybe more) to relax or to nap for the purposes of helping the firefighter shift gears from the high-octane world of firefighting to the relative chill mode of family life." When I read that column, the light bulb went on in my head: My chef husband also needed his re-entry time when he came home after a cooking gig. Anne taught me how to let him have that period when he got home to wander around the house and unwind before I approached him to discuss anything serious.

I am beyond flattered to be writing this foreword. I knew when I edited Anne's first column for Fire Life that she was destined to go on to something bigger. This book is certainly something bigger. I hope you learn from it, laugh with it, and enjoy it. If we can emulate in some small way the Gaglianos' exceptional marriage and relationship, we will all be better for it.

Diane Rothschild

Executive Editor, *Fire Engineering*,
Fire Apparatus, Emergency Equipment,
and Fire Rescue Conference Director, FDIC

ACKNOWLEDGMENTS

Anne would like to thank the following:

Diane Rothschild: My editor, creator and champion of Firelife.com, and trusted friend.

Shawn Longerich: My agent. Creative and thoughtful—you gave me the idea to write!

Bill Gustin, Jim Duffy, and Frank Ricci: You were the first big-name guys to say you actually read and liked my columns. Your encouragement inspired me to keep going.

Jonathan and Denise Jones: You believed in my speaking abilities long before I did. Your kindness and support has been a true source of strength.

Derek Rosenfeld: Thanks for all the help getting my columns and trailers posted, and for making me laugh with your sharp wit.

Bobby Halton: Thanks for taking a chance on me! And for thinking outside the box.

All our sidebar contributing authors: You made this book way better with your insightful, heartfelt stories. Thank you for having the courage to lay it all out for the benefit of others.

Paul and Sheryl Combs: Your generous donation of amazing cartoons is appreciated beyond words. Thank you for supporting us and our work, and for your genuine friendship.

John Mittendorf: Thank you for the wonderful, detailed, thoughtful foreword and for believing in marriage as we do. You are a dear brother.

Seattle Fire Chaplain Joel Ingebretson: You trumpeted our message right from the start and gave a wholehearted endorsement. You are a true champion for the firefighter's personal needs.

To all who have hosted my classes: Thank you for your vision and for seeing that firefighter marriage is elevated.

L5, D Shift; Rick Powers, Jason Kent, John Hammer, Jimmy Richards, and Mike Camlin: Thank you for always having Mike's back; I sleep better at night knowing you are by his side. Your brotherhood and hysterical, clever banter makes a tough job enjoyable. I so look forward to seeing you all every Christmas, and being first to pick!

Seattle Fire: I'm so proud and grateful to have had you as our second family for 28 years. You're the best of the best.

To firefighters everywhere: You make this world a better place. Thank you for your noble service. This book is for you, with love and deepest respect.

INTRODUCTION

The year my husband Mike made captain, he was assigned to the Joint Training Facility (JTF), first as the recruit captain then later as the training captain for the entire Seattle Fire Department (SFD). Mike had been a firefighter for nearly 15 years before deciding to promote to captain. He knew the commute would be long, as would the days if he got tagged for the 40-hour work week, which is often the case for officers. He wanted to stay on shift as long as possible to be around to raise our boys, so he waited and finally became a captain.

At the JTF he had the pleasure of running into old companions he hadn't seen for a while. Seattle Fire is a large department that encompasses a vast territory spread over numerous shifts, so it's easy to lose track of one another. But all must rotate through the JTF at some point during the year, so Mike became reacquainted with old firefighter buddies and got to meet some new ones as well. After training, many would make their way to his office to chat, to reminisce, and often just to seek counsel, as Mike has a compassionate ear. Every day he'd visit with his comrades and nearly every night he would come home to me and relate this stunning and tragic news: another Seattle firefighter was either divorced or getting divorced, some for the second or even third time.

These are good people, great people as a matter of fact, who once were happily married. The news of these wonderful families falling apart hit Mike like a truck—what was happening to them? Why were *so many* Seattle firefighters getting divorced? He asked for my thoughts as he was struggling to advise his coworkers on how to save their failing relationships. Did I, as a longtime firefighter spouse, perhaps see something that he himself did not? Are firefighters really that hard to live with? Had elements of his profession impacted *our* marriage? And was divorce just a problem for Seattle firefighters, or was it a problem everywhere?

I told Mike yes, I had a few ideas on the subject. He asked me to jot them down. I began to brainstorm why I believed this profession could be so very hard on a marriage and drew from my own decades of marital experience. Mike expected me to give him a handwritten note with a line or two. I instead handed him a 10-page, typed, annotated document. He was shocked! And impressed. He suggested I submit it to *Fire Engineering* as an article. They published it, though reluctantly. But to their surprise and ours, the response was so huge that they later offered me a regular column on their new website, Firelife.com, which began shortly thereafter. The positive feedback indicated we'd touched a nerve, an unspoken need in the fire service. All of this has inspired our book, which is written from the heart, as both Mike and I believe firefighting is one of the highest callings and that marriage is higher still.

Firefighters, you are our nation's heroes. You allow society to function in safety. Your role is crucial; your job is to save lives while risking your own. You represent the best of humanity. Without you, civilization would collapse.

Marriage—there is nothing more rewarding, fulfilling, and beautiful than a happy one. It is the most intimate relationship available to mankind. It is the ultimate source of strength, encouragement, and love, and it is meant to last a lifetime. Without love and family, there would be no civilization worth saving.

Both Mike and I have traveled and taught throughout the country for many years, and the overwhelming message from fire service leaders, chaplains, and firefighters is consistent: divorce is a significant problem in the fire service. While accurate studies are limited, the same conditions existed when Mike wrote his book *Air Management for the Fire Service*. The fire service was fully aware that cancer rates were elevated in their profession but, due to the limited and conflicting data, it could not be fully verified. Subsequent years have proven the cancer rates to be higher, and it is our contention that additional study will demonstrate the same findings for firefighter divorce rates. What is certain is the reality of the unique stressors on marriages brought by the type of work done by firefighters. This is the primary focus of the book. Chaplains have confided to me that marital and family problems are the *number-one reason* that firefighters ask for their help (this is true of police and military as well). Sadly, too many firefighter families are indeed going over the edge. The bravest and noblest among us, who need the support of marriage and family most of all, are quite likely to lose it.

Mike and I know from experience there are aspects of firefighting that have directly impacted our marriage. In this book, we identify and describe in detail how we have dealt with them. Our marriage has thus far not only survived but thrived for 30-plus years, despite the odds. Together as husband and wife, firefighter and civilian, we will give both sides of our story, with Mike weighing in toward the end of every chapter. Some of what we've learned has been by watching others, some by accident, some by trial and error, and some by sheer luck. And since we're only one voice, we've called upon other voices to broaden the scope of wisdom. Throughout the book you'll find stories written by well-respected, successful firefighter couples from across the country whose opinions on marriage we deeply value, and when it comes to marital advice, experience, in our opinion, is the very best teacher. It is our hope, our prayer, that you, the reader, will find a nugget or two of wisdom from these pages that will help your marriage. And if you do, please, please, please pass it on to a brother or sister firefighter in need.

INTRODUCTION

Fig. I–1. The day it all began for us: August 31, 1985

Part I

Three Things That Make Firefighting Different from Other Professions

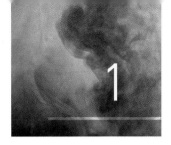

Exposure to Danger

Anne Gagliano

What is it about firefighting that makes it so very different from other professions? It is different for many reasons, some of which include uniforms, hours, and equipment. But what makes it so unique, so intense, so compelling? What separates the average office worker from the firefighter? The two primary elements that make this profession so incredibly different from others are the exposure to danger and exposure to trauma. This is no small thing, as these are two of *the* most impactful experiences to the human body and the human psyche. Firefighters are exposed to both on a regular basis. How exactly does danger affect the human body? How does exposure to trauma impact the soul? It is important to examine these experiences, as they have a direct effect on marriage. Certain behaviors arise when firefighters are exposed to danger and trauma and these behaviors aren't just displayed at the firehouse; they are also exhibited at home. To understand what is happening in firefighter marriages and why they sometimes fail, we must first look at what is happening to the firefighter.

We know that firefighters can face danger at every call. What exactly does this look like? What is danger? Danger wants to kill you; it is anything deadly. It threatens your life and your sense of well-being. It is something the average person avoids at all costs because danger causes fear—an unpleasant emotion. But when firefighters go to work, they enter the deadliest of environments. There's no cushy office routine or quiet comfort and structure. For them, it is fire out of control with heat that burns, even through bunkers. It is smoke, often a deadly black smoke in which they can neither see nor breathe. It is clouds of steam that sear exposed skin. It is impending collapse and cascading debris. And it is a deafening roar of sound as loud as thunder over which they must strain to hear the cries of both comrades and victims alike. In all of this, they must control their fear and keep their head.

Fire isn't the only danger in this profession. There's the onslaught of roaring traffic at a car wreck, the threats of unstable people, and even frightened animals who may choose to attack. Infectious blood, explosive gases, and fallen electrical wires are elements of the average workday for a firefighter. These things are deadly, these things are dangerous, and being exposed to this level of danger constantly has consequences.

FIREFIGHTERS AND PNS

The human body is controlled by the autonomic nervous system which consists of two parts: the *sympathetic nervous system* (SNS) and the *parasympathetic nervous system* (PNS). The two generally work opposite of each other. The SNS is primarily concerned with the expenditure of energy; the PNS, with building up that energy. Your brain decides, moment to moment, which is needed—energy, rest, or something in between. At night, the PNS is totally ascendant as the body goes into restoration mode. During the day, the majority of the population functions in what is known as *homeostasis*, or a perfect balance between the SNS and the PNS, where only mild amounts of energy are expended. But for those routinely exposed to danger, it is entirely different: they fluctuate between the two extremes, the result of which is accentuated highs and profound lows.

The moment the brain detects danger, a swift and miraculous change occurs. The SNS takes over, as it is responsible for the fight-or-flight response. It does this by releasing increased amounts of hormones, including adrenaline, noradrenaline, and cortisol, which prepare the body for action. These hormones increase the heart rate, respiration rate, muscular strength and skeletal tension. Cortisol helps to improve blood clotting in case of injury. The SNS releases the body's supply of stored energy while simultaneously suppressing the digestive and immune systems. The brain desires above all else to survive and will shift as much as ⅔ of its function from the cerebral cortex, or "thought" region, to the limbic or "wild-brain" region. The limbic region is essentially animal instinct—this is why training is so essential, as thinking can be impaired under duress. Shakespeare poetically describes this process: "But when the blast of war blows in our ears, then imitate the action of the tiger; stiffen the sinews, *summon up the blood.*"

The extreme high of the adrenaline rush is the fun part for firefighters. While at a fire, their SNS is ascendant and they are alive, alert, energetic, and even cheerful. A firefighter must stay alert for long periods of time, sometimes as much as 24 hours, which requires tremendous amounts of energy. By the time they get home, *PNS backlash* may hit, rendering them exhausted, apathetic, and possibly irritable. The greater the excitement at work, the greater the potential for PNS backlash that can debilitate them for family life. This must be faced and handled together if firefighter marriages are to survive.

It begins with understanding this fact: fear-induced physical exertion is profoundly different from exercise, *as hormones are involved*. Physical activity, though tiring, requires very different shifts in brain and bodily functions. Fear heightens and suppresses at the same time; exercise alone does not. Fear amplifies heart rates. Hormone-induced performance and strength increases can cause heart rates to spike to 100% of potential maximums within 10 seconds; this does not occur from even the most extreme physical exercises. Considerable amounts of glucose are also expended, including stored sources, which can cause dramatic sugar lows. After a particularly harrowing situation, a firefighter's body can crash. It must shut down to rest, to restore, to repair, and to replenish all the resources that were expended.

"Hold on," you might say, "my firefighter is not *afraid* of firefighting; he loves it." Yes, they do love it, but the brain sees fire for what it is: dangerous. Therefore the stress response automatically kicks in, whether the firefighter realizes it or not. This is a good thing, as adrenaline is required to perform at life-saving, super-human levels. In his book *On Combat*, Lieutenant Colonel Dave Grossman describes differing mental conditions soldiers go through during battle that determine whether their performance is optimal or dysfunctional. These conditions can be applied to firefighters as well when faced with life-threatening situations. The veteran firefighter typically must reach the psychological state known as Condition Red in order to survive. Condition White is where most of us live, which is the no-danger zone. Firefighters usually reside in Condition Yellow, which is a perpetual state of alert. They are ready for anything at a moment's notice; they are survivors. They have to be. But the optimum level of performance when confronted by danger is Condition Red, when heart rates spike but fear is tempered by experience, which keeps the veteran from panicking. Without experience, a firefighter can become incapacitated by fear and descend into what's known as Condition Gray or Black in which they can no longer function. Fear is a gift: it brings needed strength and courage, and with proper training, firefighters can achieve feats that seem almost impossible to the rest of us.

We marry firefighters for these reasons and more: they are strong, capable, confident, and caring. They are warriors by nature—incredibly brave and willing to fight for others—and there is nothing more appealing than that. Firefighting is sexy, but PNS backlash is not. Firefighter spouses, unfortunately, seem to mostly see the latter, as we don't get to go into the fires with them. PNS backlash is just as much a reality as PMS—both are a consequence of fluctuating hormone levels that cannot be helped. And for this reason, PNS backlash should be viewed with understanding and compassion.

CareerCast.com named firefighting as the Most Stressful Job of 2015 as it is more 'physically dangerous and psychologically taxing' than any other profession. Next on the list was the military, followed by airplane pilots, and then police. Of these top four stressful professions, firefighting is the most harrowing of all. Deputy Chief Vincent Dunn (ret.), FDNY, writes,

> The U.S. Marine Corps called several New York City fire chiefs to its military academy to find out how the fire service makes its most critical decisions. The Marines believe we make frequent life-and-death decisions during fires and other emergencies, and they wanted to know how we train our officers to do so.
>
> During a war, the military makes many life-and-death decisions, but when the war ends, so does the decision-making. The fire service, on the other hand, fights a war *that never ends.*

The Marine Corps looks to the fire service for tips! Even some of the toughest folks around realize what it means to be a firefighter: to be a warrior, *for life* (fig. 1–1).

Just as the Marines recognize the incredible stresses and strains of firefighting, so must we who love them, as this will help us live with the realities of PNS backlash. What does backlash look like? It has many faces and after nearly 30 years of this, I pretty much know them all and will go into further detail in subsequent chapters throughout the book. For the firefighter marriage to survive, we must recognize these facets of backlash for what they are—consequences of the stress response from a dangerous job, not necessarily character flaws. None of this should ever be an excuse for abuse; it is intended as an explanation for some mildly irritating or even odd behaviors that are typical to most firefighters.

Fig. 1–1. The dangerous work environment of a firefighter. Courtesy of Dennis Leger, Springfield Fire Department (MA).

Eric & Melinda Abbt, City of Houston FD, Texas

We have been married 10 years. I have been with the City of Houston Fire Department (HFD) for 14 years working on the northeast side of the city; my current rank is engineer operator. My husband Eric is a 24-year HFD veteran, and currently holds the rank of district chief on the west end of Houston. We met formally at a social event for Rodeo Safety Committee. I had no intention of ever dating a fireman. Well, I totally ate my words and ended up not only marrying a fireman, but a fireman who absolutely thought he was invincible.

Dating while working for the same fire department has its pros and cons. The pros: being able to talk shop and have empathy for the schedule and stresses of the job. The con: we often put the reality of the job on the back burner. That reality is that the dangers we face could lead to injury, illness, or even death. Eric and I never really talked about the "what-ifs" of the job. We always just believed we were exempt from all that. Well, at least he did. Eric was a seasoned firefighter who had been through countless fire incidents, and he was an instructor who owned his own fire school. He was teaching "Saving Your Own" classes, drilling with his crew, and was always prepared to handle the toughest scenes. I, on the other hand, was not as experienced; I only had a few years under my belt. Every time we left to go on shift, we would kiss goodbye and say I love you and go on our way. We used to truly believe that nothing could ever happen, until one day in March 2007.

We would often make the same fire incidents. I think he liked being with me just to make sure I was okay, and for me it was nice knowing he was near. On March 28, 2007, a few months before our wedding, Eric and I made the same mid-rise fire. I was riding Tower 18 when we were called on the second alarm. Eric, a captain on Engine 8, was called on the third alarm. When we arrived, we set up for aerial operations. I heard on the radio Engine 8 was on scene, but did not think too much about it. They had a high-pressure pumper and I had expected they would be called out. Engine 8 was ordered to go up to the fifth floor where victims were calling for help. I could hear all radio transmissions and understood they were going in. As time progressed, I started hearing radio calls for help. Still not completely aware of the situation, I just thought Engine 8 was going up to save whoever was in trouble, not realizing that it was, in fact, Eric who was calling for help!

The moment I realized the situation, my heart stopped. My captain at the time said, "He'll be okay," and ordered me down from the tower. It was a surreal feeling; I was still in disbelief that Eric was lost. I geared down and felt that if I started crying or freaking out, they would not let me see him. So I calmly asked the assistant chief, "What do I do?" He had no clue. So I decided to go to Eric's pumper, which was located on the opposite side of the building. I started to walk fast, then went into a full sprint to the other side. When I got there, I saw his crew. They were spent, and when they saw me, they had a look in their eyes I will never forget. Before I could say anything, Eric's chauffeur yelled, pointed up, and there was Eric on the tip of an aerial ladder, alive! At that moment, I realized how much this man meant to me; I had always known, subconsciously, but never to this magnitude.

Eric suffered extreme heat exhaustion, with a core temp of 107°F. I believe he had several brother guardian angels watching over him that day. He was very lucky to be alive. This was a monumental experience that really tested our relationship. My feelings during the incident went from extreme fear to instant relief. However, I did not know what was going to be in store for me after the fire.

Eric spent about three days in ICU until they felt he was well enough to leave. Though we were still uncertain whether there would be permanent damage to his body, he wanted to go home and I did too. We arrived home and slept for days, it seemed. We would face many hurdles after the fire, both physical and mental, all while the media ridiculed his actions as a captain.

Basically, our lives changed from Eric's incident. We have a strong relationship and we love each other very much. The incident opened our eyes to challenges we did not expect. I believe that every relationship that has one or both spouses in the fire service will face a situation in which they will not know what the other is thinking or feeling. I had to figure this out. I thought I knew what he was going through, but in actuality, I really didn't. We were newlyweds and I wanted to be the perfect wife and help my husband. What I realized was that I couldn't. It seemed to me that it was almost easier to deal with the actual tragic event than the aftermath of physical and mental stress. Both of us were trying to cope in different ways: I wanted to talk about it and he wanted to avoid the conversation. I did not know how to help him and I often thought to myself, Where do we go from here? This is when I realized we needed help beyond us.

Help came unexpectedly through the fire department therapist. I am not saying that every relationship needs a therapist, but sometimes you need to be open to other coping methods. Back then, firefighters were conditioned to "walk it off" and not deal with the emotions that come with a stressful incident of any kind. The therapist met with us only a few times. He came to Eric's station spontaneously, knowing that Eric would probably avoid him if he knew he was coming. To my amazement, he also came to visit me. I will forever be grateful for his words of encouragement and support. Those few visits gave us the tools we needed to open up about our relationship. We learned that Eric felt he failed me and all the while I felt I was failing him because he was still hurting. Our willingness to allow the therapist inside our world opened a line of communication I did not realize had been missing. Since then, our relationship has evolved and become stronger than ever.

One tool the therapist taught me was to understand that Eric's anger or frustration was internal and not about me. It is never you. That made sense. And I was pushing too hard to fix it instead of giving him the space he needed. I would go to him if I needed to. I did not expect him to read my mind and know what I was feeling, so why was I trying to do that to him? As a wife and a firefighter, I cannot express how important it is to step back and give them a minute to regroup. They will open up in their own time. It's almost like coming home from a long trip: when you get home, you are glad to be there, but you still need to unpack. Communication without pushiness is critical to helping your spouse decompress. And don't be afraid to use the resources around you to get another perspective on how to cope with your feelings after a particularly dangerous or traumatic run. We can't always be the hero and save the day.

If you would, please take a moment to visualize this snapshot of a firefighter in action: in the crowd of frightened faces, one young man's stands out. Vasoconstriction from the stress hormones pulsing through his veins makes the areas around his eyes, nose, and mouth a pale shade of white. Although he is dressed in the heavy protective gear of a firefighter, what distinguishes this young man from the rest of the crowd is not his uniform but one astonishing difference: he is marching up the stairs while everyone else is racing down them. This moment of both terror and incredible heroism was captured for all time by the security camera in a stairwell of the World Trade Center on September 11, 2001.

The brave young man in the photo was called and he answered. Our firefighters do this every day as they go out to meet whatever awaits them at the other end of the alarm bell. Their bodies fill with adrenaline as they "summon up the blood" to do battle for the citizens they serve. If it is an intense battle, multiple additional hormones will be used, resulting in a PNS backlash. It is in this condition that they return, God willing, to the families that love them. Stress hormones have an impact that must not be ignored by the firefighter or the firefighter's spouse, as their health and their marriage may be at stake. With great adrenaline rushes comes great exhaustion and this is what we see at home, though we don't always recognize it as such.

While PNS backlash can have many negative repercussions on both the firefighter and his or her marriage, some may find an unexpected silver lining. Yes, firefighter, when faced with death, destruction, and horror, there can be a powerful urge for greater intimacy and a life-affirming desire to reproduce. Sex is among the healthiest forms of stress relief for the firefighter couple; use it to your advantage to keep your marriage intimate and strong. Firefighter couples sacrifice so much to serve the public; enjoy some of the benefits of doing so and hold each other close.

To keep it all in perspective, always remember this: while others flee danger looking for a safe place to hide, it is the firefighter who says, "Get behind me, I will protect you." It is a dirty, dangerous, desperate job, but one that is necessary to stop death, destruction, and even terrorism. The calling to be a firefighter, to save lives, is essential if our civilization is to last. And where do we get such men and women, those willing to "ride to the sound of the guns"? We raise them up by nurturing them at home, and that makes you, firefighter spouse, even nobler still.

Chapter 1 — Exposure to Danger

Mike Gagliano

AND BY "NORMAL" YOU MEAN WHAT?

From the time I first felt the adrenaline rush of playing sports and the high that can come from listening to rock 'n' roll turned up to 10, I've felt the pull to exciting situations. The fire service allowed me to feel that excitement and channel it into a meaningful career that served others in a fulfilling and satisfying way. But it takes on a sense of normalcy the longer you do it. The stuff that makes the nonfirefighting world take a step back ends up being just another day at work for firefighters. What is critical for all who take on our calling, and surely for those who must live with them, is to recognize that these periods of danger are extreme and have a dramatic impact even if it is not recognized by those involved.

At any given moment of the day my crew and I might be called to intervene in some of the most horrific situations imaginable in an attempt to bring calm to chaos. These situations do not just occur over the course of a campaign or designated time period, but continue on throughout the course of our career, shift after shift after shift.

The exposure is ongoing, regardless of our feelings or of what is happening in our home life. It is typical for us to be washing our rigs or greeting kids at a school function only to then be cutting a bleeding patient out of the twisted remains of a vehicle. Or we might be doing a drill or inspecting a building when a ripping fire call comes in and people will live or die based on how we perform. It is rarely a "normal" day, as most would characterize normal to be.

I recall telling Anne about a run we had and the scene that presented itself on our arrival. As we pulled up, I could see an SUV on the sidewalk with a pair of legs sticking out from underneath the vehicle. A baby stroller was lying on its side next to a couple who were bleeding by a shattered store window. On closer inspection, we saw that there was a pair of legs sticking out of that broken window, as another person had been knocked through it by the car. Oh, and the driver was still in the car on top of the trapped person. It was now time to go to work.

It was only after that final bit of information that I realized that Anne was looking at me horrified. Her incomprehension of the scene I'd just painted was evident. "Normal" is a very different word away from the job. This was just work for me, nothing all that special.

This is critical for our spouses to recognize, as we become conditioned to the intensity over time and less and less surprised by what we see. It still has a dramatic impact on us physiologically, but we develop coping mechanisms, some healthy and others decidedly not. By recognizing the reality of our workplace, our hope is that you firefighter couples will determine to give each other the benefit of the doubt and become each other's champion when facing up to the realities that danger presents.

The real tragedy of not dealing appropriately with the impacts of the dangers of firefighting may very well drive you apart. Because of the bodily processes that occur to allow us to meet danger head on, some of the impacts and reactions downstream will not be pleasant when it's time to come home. The terrible reality is that so many who are giving their all at work to keep society safe and sane will lose all that is precious to them in their own homes. Since some of this is outside the firefighter's immediate control, it helps when spouses will give us some grace, see that it's not personal, and hopefully open better communication to minimize the negative behaviors and develop methods to cope with and eliminate them.

CAN HOME LIFE REALLY COMPETE?

Another challenge seems to be the mundane nature of life at home when compared to the exciting and crazy life that is the firehouse. How can changing the kids' diapers or helping with the dishes ever compare to cutting open a smashed car or opening up a roof in the middle of the night? When the "boring" normal meets the emotionally drained firefighter, it creates a recipe for argument, discontent, and lack of appreciation for home life. Both firefighter and spouse need to see this challenge for what it is and meet it head on.

The job cannot take on such a huge presence that it dominates life away from the firehouse. This takes commitment and choice. When viewed through the proper lens, a healthy family and robust marriage will enhance all that goes on while at work. The tougher runs will be filtered through a grateful heart if you have a happy home. The emotional highs and lows will be tempered by a healthy relationship that allows the venting of pain and agony, as well as the celebration of success. There is nothing that can compare to living with a spouse who is determined to be your partner and help you see the good while walking with you through the bad. You owe it to a person like that to appreciate the excitement that home life can bring and the fulfillment that is only truly found in family. Danger is real. So is the peace that comes from you and your spouse committing to being there for each other and making your marriage the top priority.

NOTES

1. William Shakespeare, *Henry V*, 3.1.5–7 (Moby Shakespeare, 1995), http://shakespeare.mit.edu/henryv/henryv.3.1.html.
2. The material in this section is derived from Dave Grossman with Loren Christensen, *On Combat: The Psychology and Physiology of Deadly Conflict in War and in Peace*, 3rd ed. (Millstadt, IL: Warrior Science Publications, 2008).
3. Vincent Dunn, *Command and Control of Fires and Emergencies* (Saddle Brook, NJ: Fire Engineering Books and Videos, 1999), 9. Emphasis added.

from emotions or even from the body.[4] ASD sufferers may perceive the world as dreamlike or unreal and may also have a poor memory of the specific events, which in its severe form is known as dissociative amnesia. ASD symptoms include generalized anxiety; hyperarousal; avoidance of situations and stimuli that are reminders of the trauma; and persistent, intrusive memories through flashbacks and dreams or visual images.

Post-traumatic stress disorder (PTSD)

If the symptoms of ASD persist for more than a month or occur more than a month following a critical event and begin to impair the sufferer's basic functions, the diagnostic terminology changes to post-traumatic stress disorder (PTSD). PTSD can happen to anyone and for numerous reasons, but it is most likely to occur when a person has experienced a traumatic event where both of the following have occurred:

- The person experienced, witnessed, or was confronted with an event in which there was the threat of actual death or serious injury. The event may also have involved a threat to the person's physical well-being or the physical well-being of another person.
- The person responded to the event with strong feelings of fear, helplessness, or horror.[5]

Firefighters witness traumatic events nearly every shift (fig. 2–1), so you'd expect PTSD rates to be extremely high. Studies vary, but their rates are much lower than they should be given the high incidence of difficult runs.[6] Why is that? Because firefighters are incredibly resilient—they have to be. Resilience is the ability to respond to adversity in such a way as to not only come through it unharmed, but to ultimately become better from the experience. Firefighters have that unique quality that allows them to face danger with courage and patience; they view a crisis as a challenge and endure harrowing life experiences with a sense of pride. They rarely respond to traumatic events with strong feelings of fear, helplessness, or horror. Instead, they rebound from misfortune and hardships and trauma with a tenacity of spirit that leaves the rest of us in awe. They accept emotional events as the cost of doing business. Firefighters spring back into shape, recover strength, and are ready to go "once more unto the breach," shift after shift, year after year.[7]

Fig. 2–1. Firefighters face potential death and trauma at every fire. Courtesy of John Odegard.

Psychologists today have changed their emphasis of study. Instead of focusing on what makes people develop conditions such as PTSD, they are investigating what makes people able to survive adversity and actually grow from it, and they are calling this the study of resilience. They most often look to firefighters as prime subjects, for the fire department has resiliency built into its very core.[8] What they are finding is that resilience isn't a quality you are born with, it's something that can be *learned*, and the fire department recognized this long before psychologists did.

Countless studies show that there are two primary reasons firefighters cope so well with trauma: one is having the front-line support of their crew, and the other is having the caring support of loved ones.[9] In this regard, firefighters are doubly blessed by having two families: the one at home and the one at the station. One helpful resource is getting counsel from senior crew members. The older ones can help the rookies cope because they've literally been there, done that. And family support is proven to be a highly effective deterrent for depressive disorders. So families, never underestimate your value.

It is believed that anywhere from 7% to 37% of firefighters will develop PTSD at some point in their career.[10] And some are more vulnerable than others. Who are the vulnerable? Estimates are that 40%–80% of people with PTSD also suffer from clinical depression. Feelings of hopelessness often cause clinically depressed firefighters to avoid talking with other firefighters after a tough run. Who is most likely to be clinically depressed? Someone whose home life has crumbled from either a death or a divorce.

A firefighter we'll call "Mickey" was found wandering in circles and crying in the parking lot of his fire station after a third consecutive traumatic run. He was going through a divorce and failed to tell his coworkers, and the cumulative stress proved to be too much for him. I know of another incidence where a firefighter's spouse was terminally ill. He chose not to tell anyone and the entire station was impacted as a result of his struggles. Are we seeing the correlation between home life and professional life? Healthy family support is crucial to a firefighter's coping ability.

Other factors that predispose firefighters to PTSD are alienation or isolation, extreme pride or perfectionism, insecurity, dissatisfaction with the job, distrust of the crew, using denial as a coping mechanism, and poor self-control. Alcohol and drug abuse are also major contributing factors to PTSD, and unfortunately, are often used in an attempt to self-medicate stress rather than seek healthier outlets.[11]

HOW THE FIRE SERVICE CAN HELP

So how does the fire department help its members (all of its members, not just the vulnerable) to be resilient? Trauma is ongoing and firefighters must face it for their entire career. The leaders must be committed to debriefing procedures that they are well prepared and well trained to carry out. This can be done in any organization, from the smallest volunteer to the biggest paid departments. It simply begins with communication. Veterans, keep your eye on the rookies. After a tough run, talk to them about it. It is normal to be emotional—tell them so. Firefighters identify incidents resulting in the death of children or fellow firefighters as among the most traumatic; be aware of these and other vulnerabilities that may be specific to your firefighters. An example of this is a barn fire I once read of. Many horses were trapped inside and the firefighters were unable to free them. They stood helplessly outside while they listened to the screams of the dying horses. A senior crew member was aware that his younger crewmate owned horses and was himself a major horse lover. He took the time to take this young man aside and let him cry privately, as he was keenly aware of the fact that this was particularly traumatic to him. Be aware of impactful runs and have a protocol in place for them. Tell your crew it's OK to cry, and if others hassle you about it, tell them to pound sand. Closely watch the firefighters who are going through difficult personal life issues (such as a divorce), as this renders them profoundly vulnerable to emotional collapse.

Resilience can be learned. Teach your people to stay calm under stress, for it's hard to be emotionally calm when you're physically aroused; emotions and actions can spiral out of control simultaneously. Type A, physically aggressive, hard-driving firefighters can learn to be Type B emotionally when handling a crisis: unruffled, relaxed, and easygoing. The best way to teach this behavioral control is to break the myths that firefighters must always be in control of both the situation and their emotions, that they must always succeed and never expose a weakness. Firefighters are take-charge problem-solvers by nature, so use this strength when discussing emotional issues: tell them the problem will be solved more quickly, head-on, if it is exposed, if they will risk expressing their feelings. Talking can help heal. Talking about a traumatic experience does help most people, most of the time, but as with everything, there are exceptions. Some people prefer not to talk about traumatic experiences at all. Compelling those individuals to talk may worsen their recovery and is a risk factor for PTSD. Find other solutions. One of these (touch) is discussed further in chapter 13.

HOW FIREFIGHTER SPOUSES CAN HELP

The same is true at home. Families can't relate professionally (unless they too are firefighters), but they can recognize odd behaviors, such as irritability and sleeplessness. Spouses, ask your firefighter what is going on if you witness unusual behavior. Firefighters, be willing to talk about it. Repressing trauma is a bad idea, since it can fester and erupt over time. No matter how tough firefighters may be, they need help battling this foe: "Unchecked extreme stress is an emotional and physical carnivore. It chews hungrily on so many of our officers with its razor-sharp fangs and does so quietly, silently in every corner of their lives. It affects their job performance, their relationships and ultimately their health."[12]

In our experience, the most typical reaction to sorrow is irritability or anger. Anger is simply the result of witnessing or experiencing a wrong. It brings a strong desire to do something about it—to fight even; thus anger is a functional emotion. It allows the firefighter to keep going, as sitting down and crying at a fire scene is simply not an option (and if a fire officer does this, they would or should be relieved of command, as lives are at stake and decisions need to be made). Anger deadens pain and provides strength to keep fighting and keep your head in the game. Anger is normal but it is also toxic to the firefighter marriage. Just as physical pain tells us to take our hand off the stove, so too does anger tell us to take heed; our psyche is being damaged. For these reasons—marital impact and underlying emotional pain—it should not be ignored.

A beloved and dear friend of ours, Captain Mike Dugan, FDNY (retired), shared with us a very personal and powerful example of this. On 9/11, FDNY lost 343 firefighters. Mike Dugan knew many of them. One was his very best friend and one his godson. He wears their commemorative bracelets to this day. Going to funerals nearly every day for weeks after 9/11 was beginning to take its toll when his youngest daughter provided a major gut-punch, a wake-up call. He realized his emotional trauma was impacting his family; they were even beginning to avoid him. He wisely sought help and he credits this with saving his marriage.

Michael & Elizabeth Dugan, FDNY

I was appointed to the New York City Fire Department (FDNY) on October 7, 1985. I had been a member of the New York Police Department (NYPD) prior to that. At that time, my future wife Elizabeth (also known as Missy) and I were dating. Missy supported me through my time in the police academy and the fire department's training academy. We were married on May 31, 1986. We have been a team through thick and thin. We've had our ups and downs, as every married couple does.

We were blessed with two wonderful daughters: Sarah was born May 6, 1992, and Emily was born February 9, 1995. During Emily's birth, there were complications and Missy was put on life support for approximately 12 days. Valentine's Day, 1995, I was called to the hospital to say goodbye to my wife because the doctors did not think she would make it through the night. Here I was with a newborn and a two-and-a-half-year-old daughter, thinking, I am going to be raising these girls by myself. During this time, the FDNY stepped up to the plate and made my life easier. I did not have to go to work for two and a half months and only went back when I chose to. They were calling me, stopping by to check on me, and bringing food. My job, the friends I worked with, and my bosses made sure I had what I needed to get through these difficult times. Missy and I made it through this together as a team. We came out stronger and more committed as a result of what we went through.

I was promoted to lieutenant in 1994 and captain in 2000. All was right with the world and our family was doing well. We had good times and we had some bad times, but through all those times we would always look for the blessings and the good to come from them. Life was good!

Then came September 11, 2001. The day before, on September 10, I was with a group of firefighters, officers, and fire department superior officers at the rededication of the firehouse where I had served as lieutenant. Ladder Company 42 and Engine Company 73 were having a party. Of the firefighters at that party, 11 did not see September 12.

I was scheduled to teach at the fire academy on September 11, 2001. We left the fire academy on a bus to respond to Ground Zero, and got there as the second tower came down. I did my time at Ground Zero on the rescue and recovery operations looking for firefighters I knew, men I had worked with, and friends who were missing. We were working 24 hours on and 24 hours off. During the day you had off, you went to funerals or services. It went on like this for a couple weeks. The FDNY was doing amazing things, they were still responding to fires throughout the city while manning the rescue and recovery at Ground Zero.

It seemed to me like we were on autopilot. We were getting things done on the job, but for a lot of us, home life suffered. Missy tried very hard to protect me and to be there for me. One beautiful day in early October she kidnapped me and told me I was not allowed to go to a funeral or memorial service but I was going with her to the beach. She had packed a lunch and we were just going to go sit on the beach and relax. I still remember that fondly.

As time went on, I became angry! I was upset with the world and pissed off at God. How could this happen? Why did this happen? I was getting more and more angry and self-medicated by drinking more than I should have. One day in November, my six-year-old daughter Emily came up and asked, "Daddy, why are you always mad?" That statement hit me harder than I have ever been hit before. One of the people I loved most in this world was afraid of me! How could this be? Missy again came through for me and said, "The kids are afraid of you because you get angry and fly off the handle." This was my epiphany, my moment of clarity, because I needed help!

I went to the counseling unit of the FDNY and told them I was having issues. It took me three tries before I found my counselor, a guy I could talk to. He was a former New York City police officer and had worked some of the same neighborhoods I had. We clicked. My wife used to call him my boyfriend. She would say, when I was getting a little testy or a little short, "Hey, when are you going to see your boyfriend?" That was her indication that I was letting my emotions take over.

Counseling and the ability to talk to someone about a traumatic event such as 9/11 probably saved my marriage and my life. I needed help; I was not able to deal with this on my own. Counseling and therapy can save marriages, families, careers, and lives. Many departments are offering these resources to their firefighters and creating critical incident stress management Teams. But because it's counseling and mental health we are afraid of it! It's time to bring mental health out of the shadows and accept that sometimes tough guys and girls do need help!

Knowing that most firefighters witness trauma, from minor car wrecks to major terrorist attacks, is half the battle. Knowing what to *do* about it, how to live with it, and how to actually help your spouse is the other half. And how do we help? What are some of the best ways to safeguard the firefighter marriage against the ravages of trauma? This topic will be detailed more specifically in chapter 6 but here are some basic principles to start with.

Take time

He that lacks time to mourn, lacks time to mend.

—*Sir Henry Taylor*

Time off is essential for the traumatized; encourage your firefighter to take it. Returning to work too soon could result in another incident for which the firefighter is not emotionally prepared. Time with you is healing—time for sex, time for laughter, time for singing as "these are powerful survival mechanisms that have been developed across the millennia to help defuse traumatic situations by reasserting normality into our lives."[13] Family normalcy, routine, and rest are the quickest ways to soothe a troubled heart. Firefighters often have generous sick-leave. Use it wisely to play the long game, both professionally and relationally.

Understanding

Understanding is essential to the firefighter marriage. Understanding is not empathy but a comprehension of this one fact: we can never fully grasp what our firefighters have been through, even if we are firefighters ourselves. A huge chasm exists between what they've experienced and what we have, and would even if we'd been there. This gap will always exist, so just accept it. Don't press for details that may be too hard to relate and don't tell them how they should feel when you clearly cannot know. What they need to hear first and foremost is that you're glad they're home safe, that you're there for them, and that you love them no matter what.

Be OK with the fact that they may need to share some of the details with a coworker instead of you. In fact, encourage this. Numerous firefighters have relayed to us that debriefing with coworkers and/or critical incident stress teams is both healing and healthy. Rehashing the experience with those who were there helps firefighters make peace with the memories, which alleviates avoidance behaviors such as anger. Many fire departments are implementing this practice and getting great results. Mike has relayed to me the great benefits being seen in Seattle Fire and I've even participated in some of these myself.

Affirmation

The traumatized are very vulnerable to guilt, shame, even self-loathing. Strong, confident firefighters may misconstrue feelings of sorrow as weakness, even failure. Your loving presence powerfully refutes these lies. Tell them that you are proud of them, even if they made mistakes. Your unwavering belief in them will help restore their shaken confidence.

Support

True support means saying "thank you" for all that they do. It is appreciation for a tough job that most people would not or could not do. And when it comes from a spouse, it restores and revitalizes like nothing else. Support literally means to help stand, giving the firefighter the strength to endure incredible adversity.

When a firefighter has been traumatized, their psyche is altered and can be temporarily susceptible to outside influence for good or for bad, according to Acosta and Prager. They write, "An altered state is like fertile soil....We can either say and do nothing, use our words and our presence to heal, or use words to harm."[14] Have a plan in place ahead of time, firefighter couples, to be ready when that traumatic event occurs. We believe mental preparation will dramatically decrease emotional impact on the firefighter and ultimately on your relationship. Having *no* plan increases the likelihood of PTSD and marital implosion and we have seen this with numerous couples. Our feeling is that marriage has a much more significant impact than other relationships in its ability to heal or to harm. In the movie *Lincoln*, directed by Steven Spielberg (2012), Abraham Lincoln says this best when speaking to his wife Mary while leading our country through the horrors of the Civil War: "What I carry within me—you must allow me to do it, alone as I must. And you alone, Mary, you alone may lighten this burden or render it intolerable. As you choose." What will you choose?

With the support of two families, firefighter, you can struggle well, face adversity with success, and overcome any obstacle with confidence. Take as much care of yourself emotionally as you do your equipment and you will be strengthened by adversity, not destroyed by it.

| Chapter 2 | Exposure to Trauma |

Every life, at some points, must be tested by adversity. We must face losses we can neither prevent nor reverse, confront threats we can never fully neutralize, and master challenges both sought and assumed. With adversity come sorrow and distress, but from its mastery come strength, character, and resolve.

—Richard Gist

TRAUMA IS VERY REAL

The idea that we can be traumatized seems so dramatic. Many of my cohorts would scoff at the notion that something they have seen has "traumatized" them. There is real danger in that attitude. Just as the physical aspects of the job must ultimately be performed by flesh and blood human beings, not robots, so too, the mental and psychological aspects of our profession are being felt by human beings who react in typical ways to witnessing terrible things. It is a real indication of the progress within the fire service that most everyone now recognizes the impact of seeing trauma and dealing with its outcomes. Especially as we are called to these runs more and more all time, much more so even than actual fires.

On more than one occasion I've brought the lingering effects of bad runs home to my family. It normally manifests itself in being grouchy, detached, or distant. While I am relieved to say it has never crossed over into any type of physical abuse, I'm not proud of some of the things that have come out of my mouth, or the days I've ruined by not dealing honestly and effectively with the stuff that is going on inside. Every one of us knows how real it is. It may affect some in more extreme ways than others, but the impacts are at times overwhelming.

The stacking of bad runs seems to be what gets me as much as anything else. If they are spread apart there is an obvious time element that allows you to process the horrors and to deal with them using varied means. When they hit one after another, combined with lack of sleep, the darker side seems to have a field day. I do desperately need my wife when those times come.

If only every firefighter spouse recognized the key role they play. You are such an important part of completing the mission and helping so many who cannot help themselves. Your compassion for a deeply saddened firefighter will ease their pain. Your understanding when your firefighter comes home from the firehouse and is not himself or herself is truly heroic. Giving the benefit of the doubt to the grumbly spouse who is not going with the plan of the day is just as important as the plan itself.

BUT WHAT ABOUT ME?

I know some spouses reading this will immediately say, "Hey, I've got a tough job too, what about me?" That is the mantra of our current time: "What about me?" We're advocating for behaviors and commitments that move a marriage toward contentment and mutually beneficial goals. Many of the concepts in this book apply equally to how firefighters can help their spouses. I cannot begin to count the many times when Anne was patient with me, encouraging and supportive. In those situations, the unselfishness of her actions, naturally, caused me in turn to respond with extra love, attention, and support. I need to model that as well to have the marriage we want.

It is critical that firefighter spouses remember they are sending the most important person in their life off to a world that is anything but normal. It is hard to describe the sweep of emotions that encompass a firefighter's day. We literally show up for work with one ear attuned to a radio while we go about the necessary relief of our fellow members. At any time during the changing out of gear, exchanging radio batteries, signing onto the computer, getting updates from the relief crew, getting the day's schedule, joking around with fellow firefighters, and answering the various phone calls, we could get an alarm and go from normal routine to 100 mph with the ring of a bell. And we're off. It doesn't matter if we're in the middle of using the restroom, cooking our morning meal, filling out reports or chatting on the phone: we roll. The run has our full attention, our full intensity, our full strength.

And then it's over. It could be a cut finger, some smoldering grass, a drug overdose, a SIDS (sudden infant death syndrome) death, a flooded basement, a house fire, a heart attack, a car accident, an anxiety attack, a downed powerline, a building collapse, or countless other situations for which there is no other solution than calling 911. We thrive on meeting those needs, but they create needs for us that even we find hard to explain.

YOU PLAY THE CRITICAL ROLE

Helping us deal with it and seeking compassionate, creative ways to meet those needs is one sacred blessing only a firefighter spouse can give their mate. If taken as a burden, you'll likely burn out or feel it shouldn't be your job. If taken as a challenge and a privilege of being married, it can be an enormous accomplishment that will result in wonderful returns. Firefighters are capable of incredible devotion and commitment. Our job requires that of us. Imagine that devotion turned full-on toward the person who is always there, always in our corner, always our dearest friend. With that type of teamwork, trauma and its effects stand no chance.

NOTES

1. Parts of this chapter were originally published in a slightly different form in "Resilience," *Fire Life*, January 7, 2013, http://www.fireengineering.com/articles/fire_life/articles/2013/january/resilience.html. Reprinted by permission of the author and the publisher.
2. *Merriam-Webster*, s.v. "trauma (*n.*)," updated December 18, 2017, https://www.merriam-webster.com/dictionary/trauma.
3. Monash University, "Stress Reactions to Trauma/Death." http://www.adm.monash.edu.au/community-services/crisis/stress-reactions.html, 2007.
4. Laura Gibson, "Acute Stress Disorder," National Center for PTSD, February 23, 2016, https://www.ptsd.va.gov/professional/treatment/early/acute-stress-disorder.asp.
5. Matthew Tull, "Rates of PTSD in Firefighters," Verywell.com, February 15, 2017, https://www.verywell.com/rates-of-ptsd-in-firefighters-2797428.
6. Ibid.
7. Shakespeare, *Henry V*, 3.1.1.
8. Ellen Kirschman, *I Love a Fire Fighter: What the Family Needs to Know* (New York: Guilford Press, 2004), 175.
9. Ibid.
10. Tull, "Rates of PTSD in Firefighters."
11. "The Link between PTSD and Substance Abuse/Addiction," American Addiction Centers, https://americanaddictioncenters.org/ptsd/.
12. Grossman and Christensen, *On Combat*, 4.
13. Ben Shephard, A War of Nerves: Soldiers and Psychiatrists in the Twentieth Century (Cambridge, MA: Harvard University Press, 2003).
14. Judith Acosta and Judith Simon Prager, *The Worst Is Over: What to Say When Every Moment Counts*, rev. ed. (N.p.: CreateSpace Independent Publishing Platform, 2014).

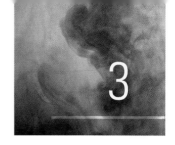

THE 24-HOUR SHIFT

Sleep disruption is another form of trauma for the firefighter. On a 24-hour shift, a firefighter may get to sleep at some point during the night. But this is sleep in name only, as at any time, they may be awakened suddenly and abruptly by a bunkroom light and some form of a bell indicating there is an alarm. This is shocking to say the least. For the on-call firefighter at home, the disruption from a radio or pager in the night has the same affect. Bobby Halton, editor in chief of *Fire Engineering* magazine, once wrote of a study that discovered a firefighter's heart rate typically spikes to 80% of maximum when the alarm is received. This is a disturbing number, as 80% of maximum is rarely achieved even in strenuous exercise.[1] Not only are they roused out of sleep, but the firefighter must then quickly don bunker gear, race to the apparatus, and receive instructions, all in a matter of minutes. The heart rate spikes for this very reason, as it must provide the additional blood flow needed to transform the body from complete rest to complete action. The 24-hour shift is an unusual work schedule, another element that makes this profession unique. It is both a blessing and a curse, rewarding, yet consequential. And it is traumatic, as all shocks to the system typically are. The long-term effects are something firefighters and their spouses should be aware of.

Why the 24-hour shift anyway? How did it start, where did the idea come from, and why don't other professions operate this way? The 24-hour shift dates back to the 1800s and the birth of big-city fire departments in the United States. The early firefighters didn't work shifts at the firehouse, they *lived* there. It was a new profession that wasn't overly stable, profitable, or respected. For these reasons it drew single young men who could survive on very low wages while also providing them with a place to live. And as the rigs were drawn by horses back then, someone needed to care for the animals around the clock as well.

By the early 1900s, as populations grew, so did the need and respect for the fire department. The profession was recognized as honorable, life-saving, and worthy of full support from the communities. Born of a noble desire to preserve and protect human life, the commitment to providing 24-hour protection, 7 days a week arose. To accommodate, firehouses sprang up all over the cities, strategically placed to reach every corner at every hour as quickly as possible. With more buildings came more firefighters, and as the need grew, so did the appreciation—and the wages. Higher salaries drew men who could actually support a family, and a family man couldn't live at the firehouse, so they requested a day off for every day worked to be with their families.

Why don't cops work 24 hours, since they also provide round-the-clock protection? The difference is this: they must stay awake their entire shift and actively patrol. It is a vigil that is *proactive* or *protective*. A presence on the streets. Firefighting, on the other hand, is *reactive* or *responsive*. They wait for the call. It is unnecessary to drive the streets looking for fires. And now, most firefighters provide emergency medical services as well (as they typically can get there more quickly than a hospital ambulance). Firefighters take reactive measures, they don't drive around preventing fires or accidents the way police try to prevent crimes. Because firefighting is responsive, the waiting can be done at the fire house, where they *might* sleep, but are never *guaranteed* to.

Critics have argued the need for the 24-hour shift, asking whether a firefighter should actually earn a wage while eating, grocery shopping, even sleeping. The research on the cost effectiveness of the 24-hour shift has surprised even the harshest critics. It is not a tax-payer loss, but a tax-payer *gain*, saving considerable amounts of money. To move to 8- or 12-hour shifts would require double, even triple staffing, and much more overtime pay.

Arizona State University (ASU) did a study on cities that strive to adequately staff their fire departments. Critics claim that paid firefighters are not worth the 24-hour protection that comes at the taxpayer expense, but the ASU study proves otherwise. It found that in a three-month period dating from June 1 to August 12, 2012, the Phoenix Fire Department saved an estimated 2,300 jobs from being eliminated and $10.6 million in state tax revenue after the cost of the firefighters' salaries was deducted. $10.6 million in three months? Firefighters, I'd say you earn your keep and then some.[2]

And what critics would likely find most surprising of all is the fact that firefighters *don't* rest. They don't just sit around and watch TV while getting paid to do so. Today's firefighters have effectively evolved into specialized, multirole responders who meet the diverse needs of citizens when they are at their most vulnerable. Their role is ever expanding as society continues to heap massive expectations upon their shoulders. Today, most firefighters are trained for emergency medicine, building collapse, search and rescue, high-angle rescue, hazardous materials incidents, automobile accidents, water rescue, wildland fires, infectious disease, domestic terrorism, weapons of mass destruction, and active shooter incidents. Oh yeah, and they fight fires too, over a million a year.

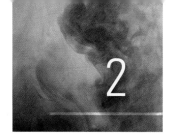

Exposure to Trauma

Are firefighters traumatized? This is an important question, as exposure to trauma is the most impactful experience known to the human psyche. Obviously, if someone is traumatized, it can dramatically impact their marriage. For this reason, spouse, it's important for you to know if and when your firefighter experiences the following:

- A shooting or stabbing
- Someone getting hurt or killed
- A horrific traffic accident
- A dead child or infant
- Inability to rescue someone
- Loss of a coworker
- A debilitating or life-threatening injury[1]

If your firefighter has endured any of the above, they have likely been traumatized (firefighters report that the death of a child or a coworker as being the most traumatic of all). *Trauma* is the Greek word for "wound," in this context a wounded heart.[2] Anyone with a heart is susceptible to being hurt, and let's face it—firefighters are *all* heart. When exposed to critical incidents, it is both typical and normal to have physical and emotional reactions. A firefighter spouse should be on the lookout for certain behaviors at home.

SYMPTOMS OF TRAUMA

Psychological and emotional

- Heightened anxiety or fear about the death of others, about the future
- Irritability, restlessness, overexcitability
- Feelings of sadness, moodiness, more crying than usual

- Feelings of numbness or detachment
- "Survivor guilt" or feelings of self-blame
- Mood swings; small reminders or seemingly insignificant emotional events trigger sudden changes in mood or intense reactions

Cognitive

- Difficulty concentrating
- Feeling confused, disoriented, distracted, unable to think as quickly or as easily as usual
- Difficulty making decisions that normally would be easy
- Worrying about death or thinking constantly about people who have died

Physical

- Headaches
- Nausea or upset stomach
- Exaggerated startle response (easily startled)
- Fatigue; a lot of energy goes into grief, and it can be overwhelming and physically draining

Behavioral

- Hyperactivity or less activity than usual
- Withdrawal, social isolation
- Avoidance of activities or places that bring memories of the event
- Loss of appetite
- Inability to fall asleep or remain asleep, disrupted sleep, deep sadness on awakening[3]

STRESS DISORDERS

Acute stress disorder (ASD)

On occasion, these symptoms can descend into a slightly more severe state called acute stress disorder (ASD). In ASD, a person may experience "numbing, reduced awareness, depersonalization, derealization, or amnesia," all of which are symptoms of dissociation, or the perception of a detachment of the mind

Chapter 3   The 24-Hour Shift

Citizens continually expect more and more of firefighters while giving them less and less, and yet they graciously comply, without ever going on strike to demand higher wages for their continuously increasing workload.

Cardiopulmonary resuscitation (CPR) and defibrillation administered within four minutes of the event dramatically increase the chances of survival for the victims of cardiac arrest. Because fire stations are located throughout communities, firefighters can reach far more citizens much quicker than hospitals ever could. From Seattle, with its average annual EMS run load of 97,000[3] to New York, where the load is approximately 1.7 million a year,[4] people are consistently reached within the critical four minutes. Firefighters save cardiac arrest victims 24 hours a day, 7 days a week. What price can you put on that?

Survival for critical trauma patients depends on prompt response and rapid transport to a trauma center. From car wrecks to shootings, firefighters are there, securing and transporting fragile lives within the crucial timeframe to receive life-saving surgery. Their skill, speed, and heroism turn tragedy into miracles in the light of day or in the deepest darkness of night.

The community reaps huge benefits from the 24-hour schedule, and in many ways, so does the firefighter. It is, hands down, our preferred schedule.

Sean & Adrian Gray, Cobb County FD, Georgia

- Marry someone with the same energy level as you.
- Respect each other's time; a little quality time goes a long way.
- You don't have to be the best at everything, but you have to make the most of everything.
- Love your job, but love your family more.

Just some background on us. We were both married young and for a very short time to other people prior to us meeting at work. We were going through divorces when we met. So, since we were technically a rebound relationship we shouldn't have worked out, right? Instead, our personalities were a breath of fresh air to each other and I believe our relationship today is successful because of our similar personalities.

We're both high-energy, high-achieving individuals. We both excel in our jobs because of our driven personalities. However, Adrian doesn't ever want to be more than an ER nurse. She's not interested in being a supervisor or going to school to be a nurse practitioner. She tried working as a flight nurse for a year and hated the egos involved, so she quit. She loves her job working in a Level 1 trauma center in downtown Atlanta. She's been there for 17 years and wouldn't change it for anything. I know—she sounds like the crusty old firefighter who just wants to ride back step. She's very good at her job, though she would never say it. Many of the younger nurses reach out to her for advice and she's the nurse all the doctors want next to them when things get bad. Being driven to be good at our jobs is something we have in common. This helps us as a couple because when one of us digs in at work we both understand. Adrian enjoys her time off and being alone in the woods on her bike. I enjoy being with her and the kids. I don't need a lot of alone time, as I thrive being around people.

Firefighter marriages are unique. We've had our struggles and our current situation is not easy. A 40-hour work week is not what it's made out to be. Adrian and I do better when we see more of each other. Being home together at night seems like it would be better for a marriage, but it's not! Spending time together during the day while the kids are at school is the best thing for our marriage. Also, Adrian enjoys her time alone when I'm at the station for a 24-hour shift. Firefighter marriages are unique and I couldn't imagine being with anyone else but Adrian. We plan to grow old together even though we can't stand each other at times.

REASONS WE LOVE THE 24-HOUR SHIFT

When others run away, firefighters go forward, providing reliable, efficient, and effective all-hazards response clearly seen throughout the country day after day, year after year. In the darkest hours of America's most dramatic events, such as 9/11, Hurricane Katrina, Superstorm Sandy, civil unrest, shootings, and terrorist attacks throughout the land, firefighters are there. They face flooding, tornadoes, wild fires, and even massive mudslides in places like Oso (WA).[5] On our worst days, they give us comfort and hope and life, and they do this through the night, sleeping with one eye open. This schedule, this life, is a blessing to us all, providing service at a fraction of what it would cost the private sector to duplicate. Emergencies come at all hours; therefore, our responders must work all hours.

The 24-hour shift has many positive advantages, not only for the community, but for the firefighter as well. In the big city, it means less commuting—*much* less commuting—which results in more time at home. It also provides an easier transition from home life to fire life. A day for all of us is 24 hours; it's easier to stay in one mode for that 24 hours than it is to shift rapidly back and forth.

Easier for the firefighter and easier for the spouse (I enjoy my occasional evenings to myself).

Firefighting is a team occupation, one that cannot be accomplished alone. The 24-hour shift gives firefighters a unique opportunity to bond, not only as a team, but as a family. Families connect over routines such as planning and sharing meals, and joking about various rituals and peculiar habits—like snoring. And the family bond is sacred, as they will risk all for one another. When firefighters live together, they become as family with the deepest levels of devotion. This devotion makes firefighters safer and better at what they do.

The meals that firefighters share are typically fresh and cost-effective, as most homemade meals are. This too is beneficial and more fun (fig. 3–1). Night-shift only or day-shift only workers tend to eat fast junk food, as they have no time for well-prepared meals, and they often eat alone. This type of eating can be expensive and unhealthy, as highly processed foods (or quick foods) are full of preservatives and lack nutrition. And let's face it, eating alone is not as pleasurable or relaxing as breaking bread with a friend.

Fig. 3–1. The "Beanery" at Station 31. Members enjoying a meal and the best fellowship in the world during a 24-hour shift.

My firefighter and I have had the pleasure of living with the 24-hour shift for most of his career. We prefer it, hands down, to any other schedule. When Mike made captain, he was assigned to the Joint Training Facility for two years as the training officer. He now had to do the 40-hour work week. He spent nearly 4 hours a day stuck in soul-stealing commuter traffic. We lost so much time together! We were relieved when this assignment ended and he went back to the 24-hour shift. He works longer chunks, but he's home longer too—we've had the privilege of having him around, and we, his family, have loved it.

The blessings of a 24-hour shift include cost-effective, around-the clock service to the community, tighter bonds with the crew, less time shifting between worlds, and more precious, invaluable time at home. These are the pros, the major perks. But there is, however, a hefty cost for this preferred schedule, one that can be quite harmful to the firefighter's health and relationships: circadian rhythm disruption.

CIRCADIAN RHYTHM DISRUPTION

"Circa" is Latin for "around" and "dian" is Latin for "day," so a circadian rhythm is the body's internal, "around the day," 24-hour clock. It is a major body rhythm with regular ups and downs that cause many systems in the body to be active at certain times and inactive at all other times. Usually, body systems are most active in the late afternoon or early evening; for example, the body's ability to produce energy from food (metabolism) is highest between 4 and 6 PM. The least body activity usually happens in the middle of the night when most people are sleeping, between 4 and 6 a.m. Circadian rhythm affects how alert people feel. The ability to perform work is best when the rhythm is high and worst when it is low. The rhythm prepares your body by releasing chemicals needed to be wakeful in the day and sleepy at night. Circadian rhythm disruption is when the body's clock is thrown off from its normal patterns, which confuses body functions such as digestion and the wake-sleep cycle. This disruption can be moderate, as in jet lag, or it can be severe, as in being awakened in the night by loud bells and whistles, night after night, year after year—the career of a firefighter. This is why this topic is serious and one that firefighter couples need to be aware of.[6]

Chronic circadian rhythm disruption is much more than just lack of sleep. Those who experience it are 60% more likely to suffer from the following:

- Obesity
- Cardiovascular disease
- Mood swings
- Gastrointestinal problems
- Motor vehicle crashes
- In persons with diabetes, trouble controlling blood sugar levels
- Low testosterone levels or hormonal imbalances

- More sensitivity to toxic exposures
- Substance abuse
- Chronic fatigue
- Family problems and divorce[7]

The three primary synchronizing agents of circadian rhythm are light, social and physical activity cues, and melatonin.[8] Of these three, light is the strongest. Light affects melatonin production, which is the chemical produced by the brain that causes sleepiness. This is why night work can be harmful: darkness is tough to duplicate in the day, so catching up on sleep for the nightshift worker is often hard to do. Light simply means wakefulness to the human brain; it's *very* difficult to sleep when any light is present. The central circadian rhythm pacemaker in humans is located in the suprachiasmatic nucleus (SCN) area of the brain, which regulates the release of melatonin from the pineal gland, which is supposed to begin about two hours after sunset. Light suppresses this secretion (and so does the stress hormone cortisol). Over time, an inadequacy of melatonin can lead to insomnia, or the inability to fall asleep naturally, and this can lead to substance abuse.

The SCN is also responsible for regulating many body functions that revolve around the 24-hour clock, including body temperature, heart rate, blood pressure, metabolism, bowel movements, and the release of hormones.[9] When this clock is disrupted, *all* of these functions are interrupted, causing irregularities and imbalances. Testosterone is produced during REM sleep; without adequate sleep, testosterone levels drop (cortisol may also suppress testosterone levels.) Low testosterone causes fatigue, moodiness, and decreased libido, all of which can impact a relationship.

These are some of the potential hazards of the 24-hour shift and its impact on circadian rhythm. And let me stress that these symptoms are *potential*, they are not absolutely guaranteed. We have options, firefighter couples, and if we choose to use them, our firefighters can be healthier and so can our relationships. There are ways to help your firefighter reset the circadian rhythm and keep the disruption to a minimum. These are nice to know, since many of us prefer the 24-hour schedule to any other. Have a plan in place so that the impact of circadian rhythm disruption will be minimal to both your firefighter and to you.

How to regulate circadian rhythm

In order to compensate for circadian rhythm disruption, it helps to know what regulates it. Remember, the three primary synchronizing agents of circadian rhythm are light, social and physical activity cues, and melatonin. Specialists focus on these three when treating sleep disorders, as they can be used to help reset or restore circadian rhythm once it has been disrupted. The more we firefighter couples know about circadian rhythm, the better we will handle this schedule long term. Over time, we too can become "specialists" of a sort as we learn to help our firefighters with the repercussions of the 24-hour shift.[10]

Light. Of the three synchronizing agents, light is the strongest. It plays a crucial role in helping our bodies identify the time of day and the levels of energy they should produce. Light aligns circadian rhythm and exposure to strong light during the day is essential for increased energy and levels of productivity. Blue light is the strongest, white light is the second strongest. But the problem with strong light, particularly blue, is that exposure to it at night interferes with the ability to sleep, as it impairs the pineal gland's ability to produce melatonin, the sleep hormone.

For the firefighter, this is important. Over time, being awakened with light renders them hypervigilant, or even more affected by light than the general population. Blue light is the strongest, so it is where we should start. Avoid blue light at least one, but preferably two to three hours, before bed. This will help anyone fall asleep faster, but for the hypervigilant, it is especially helpful. Sources of blue light include LED lights, TV screens, computer screens, and cell phones. Applications that will dim your electronic device's blue light or even change it to a less-impactful white light are available for free download. Some apps even self-regulate by mocking day patterns and gradually dimming as evening approaches. Blue-blocking glasses are also an option that can be worn in the evening.[11]

Studies on red lights are emerging, and this is an interesting option for the circadian-disrupted firefighter who has trouble sleeping. Red is the lowest energy, slowest moving form of visible light that has the least impact on the eyes. Red tricks the body into thinking no light is present and in turn encourages the release of melatonin. Red light apps, too, are available as downloads. Putting a red bulb in the bedside lamp, using them as night lights in the bathroom, or even putting them in fire station bunkrooms is food for thought.[12]

In the firefighter bedroom, light-suppressing curtains will help, as even the dimmest of lights can impede melatonin secretion. Street lights, car lights, sunlight—all of these can shine through and awaken the easily awakened. Darkness is the best sleep aid and the quickest way to reset circadian rhythm. It is terribly difficult to find in this modern age of 24-hour electronic availability, but firefighter couples should make every effort to supply it as sleep is such a precious commodity for the night-shift worker.

Melatonin. Light suppresses melatonin and studies show that the sleep-deprived have lower levels of melatonin than the general population.[13] So it is here, firefighter families, that we have another chance to exhibit some control: find sources of melatonin to give to your firefighter. Without normal levels of melatonin, insomnia can develop, or the inability to fall asleep naturally. For the circadian-rhythm disrupted, this is a common problem as the wake-sleep cycle is off and the body has trouble knowing when to produce and release melatonin. In desperation, insomniacs often seek sleep aids such as alcohol or pills or both. These tend to be unhealthy options, and they can also have serious repercussions for the firefighter marriage. Melatonin is a natural, healthy alternative with zero relationship side-effects.

To jumpstart the severely deprived, melatonin supplements are an option, but dosage levels vary widely. I prefer natural sources (i.e., food). The very best source of naturally occurring melatonin I have found is in tart cherries. And they have to be tart: sweet cherries have 50% less melatonin and dried cherries have none at all. A little glass of Montmorency cherry juice before bed has been shown to raise melatonin levels and thus promote sleep. (If you have trouble finding it in your store, try buying it online.) There is an anti-inflammatory property in tart cherry juice that also makes it such an excellent sleep aid, as this, too, aids in relaxation. Other natural sources of melatonin include pineapple, bananas, oranges, oats, sweet corn, rice, tomatoes, and barley.[14]

Social and physical activity cues. I believe this to be the primary reason firefighters have trouble catching up on sleep and restoring their circadian rhythms after a 24-hour shift. Social and physical activity cues include a desire to be a part of family life and the pressures to perform daily tasks as everyone else does. Most night workers prefer to join in the family activities over sleeping during the day. Or they may feel compelled to work a second job on their days off, as many feel like society dictates that if you don't work every day, you're lazy. On average, night workers get two to three fewer hours of sleep per night than the general population, and the desire to be with family is the number-one reason given for this (more work is number two).[15]

To be healthy, adults need 7 to 9 hours of uninterrupted sleep; firefighters typically get 5 or 6, even on nights spent at home, because of an inability to fall asleep naturally. Naps can help, but effective length will vary based on the individual and level of fatigue. It is during REM that hormones are produced, as this is when the endocrine system comes to life. Testosterone levels are influenced greatly by sleep: 4 hours of sleep produces 200–300 nanograms per deciliter (ng/dL), whereas eight hours produces 500–700 ng/dL—that's a 60% difference. Over time, low testosterone levels greatly impact a man's energy levels, ability to focus, mood, and libido. Low testosterone is also linked to snoring and sleep apnea.[16]

To help your firefighter recover more quickly from the ravages of night work, encourage long daytime naps by being flexible about family involvement whenever possible. It is better to miss out a bit here and there to restore mood and optimal health. Better for the firefighter, and better for the family, so, firefighters, try not to feel too guilty about it. And firefighter couples, maybe it's more fiscally prudent if they don't take on that second job; too much work and not enough sleep may end up costing your family more than the loss of an extra income. One job really is enough, particularly one that's done well over the long haul. And try to keep the honey-do list the same length as anyone else's—just because a firefighter has free days doesn't mean they should be overloaded.

Some daytime sleeping tips include the following: Have a quiet room with light-blocking curtains away from the main rooms of the house (kitchen or other family centers). The kids must be allowed to play; keeping them quiet is tough, as I well know from personal experience. But if you have a quiet day-sleep room, this will help keep everyone in the family happy: it lets the sleep-deprived firefighter catch up while still allowing the kids to be kids. Try to avoid loud chores, turn down the TV, and disable the doorbell. Remove the phone from the sleeping room or turn it off altogether; a few missed calls will be worth it to restore a firefighter's mood after a long night.

Try to keep daily routines as regular as possible. Go to bed at regular times and perform habitual nighttime routines (such as brushing teeth or bathing or reading), as these will tell the body when to start releasing melatonin. Also, avoid heavy foods and alcohol before bed. Light snacks actually promote sleep (such as cheese and crackers), but heavy greasy foods require the stomach to work harder and this inhibits sleep. Avoid intense exercise three hours before bed as this too can wake you up. And it is best to sleep cold and naked: a cool room and loose or no clothing helps restore body temperature and blood pressure, which are thrown off by circadian rhythm disruption.

Firefighter family, be sensitive to the fact that your firefighter needs rest after a 24-hour shift. Science has shown that two nights of uninterrupted sleep is the minimum needed to reset the circadian rhythm after just one night of disruption, with four nights of uninterrupted sleep being the optimum.[17] Those nights off are given for a reason; they help firefighters play the long game by giving circadian rhythm a chance to reset, thus avoiding chronic conditions such as digestive disorders. The 24-hour schedule is better than rotating 12-hour shifts, as it gives the body a chance to stick to a typical day/night pattern that is more in synch with family life. Use those off days to their full potential by staying on as regular a schedule as possible. Avoid too much extra work and other irregularities, such as all-night movie and pizza binges, as these will keep your firefighter's already-disrupted body clock perpetually off balance.

Utilizing the properties of light and dark in such ways as providing dark rooms in which to sleep and switching to more red light and less blue is step one in resetting disrupted circadian rhythms. Enhancing the diet with foods rich in melatonin is step two. Step three, social and physical activity cues, requires the patience, understanding, and support of the firefighter's family: patience when a daytime nap forces the family to be quiet, understanding of the physical and emotional tolls of being up all night, and support when the firefighter struggles to choose sleep over a family activity or extra work.

With these three tools, we firefighter families can help our firefighters keep circadian rhythm disruptions to a minimum. Sleep is the firefighter's friend. With sleep comes the restoration of balance to all systems and this balance will improve health, mood, and family life. An effort to reset the body clock as quickly as possible gives the firefighter the best possible chance of having long-term success in both arenas; firefighting *and* family.

SLEEP, CURSED SLEEP

Sleep has always been elusive for me, even back when I was a kid. I suppose becoming a firefighter was fortuitous since I can't really sleep that well anyway. Make no mistake, sleep is a problem for all firefighters. Whether they are paid firefighters on a schedule that has them working a 24-, 48-, or even 96-hour shifts, or our many volunteers or paid on-call firefighters who are roused from their homes, sleep disruption is a well-known companion.

It is always fascinating to talk with my friends who've retired from the job. While the details vary and some miss it more than others, one thing is always the same: they love being able to finally sleep. The ballpark time frame seems to be around three months away from the job before a new routine starts to happen. Most say something along the lines of "I never knew how tired I was until I actually started getting rest again." This is certainly true as we age in the profession.

Make no mistake about it: I love the 24-hour shift and all the benefits it provides my family. I am grateful for it and glad that it is also economically positive for the city and the citizens. The ability to really bond with my crew and develop a camaraderie that is essential for our mission is something I'll always be thankful for. All good things have downsides, however, and for the firefighter marriage, the downsides of the 24-hour shift end up being sleep and being gone for extended periods of time.

Since sleep is handled at length in a future chapter, I'll focus on some of the challenges we've faced being apart for longer periods than is normal in most jobs.

GOING, GOING, GONE

One thing to get settled in your mind right now is simply this: if something is going to break at your house, it will happen while you're on shift. That is a universal truth as reliable as gravity. The kids will get hurt at school, the car will not start, and your neighbor will pick that time to cut down a tree that drops in your backyard. It is as predictable as rain in the spring.

Firefighters and their spouses need to develop contingency plans for the things that will come up during these absences. It will save a ton of stress and more than a few angry phone conversations when the meltdown at home cannot be addressed quickly by the spouse at the firehouse. It may sound simple, but this problem has caused some terrible divisions in relationships we know. For example, one friend of mine lost his wife to a neighbor who was conveniently always available to help out when trouble happened. The gaps in being home are inevitable, but if plans are devised by both partners, it alleviates aggravation and allows for contingencies that will not send things sideways.

A BREAK TO NOT BRAKE

Another aspect of the 24-hour shift that is critical to understand is that we will have extra time off. That time off is needed for recuperation. Not all of it, obviously, but certainly some. And the temptation (for the firefighter or the spouse) to fill that time with extra work, extra play, extra chores, extra, extra, extra can lead to a hectic, exhausted lifestyle. And it will not matter if the firefighter is blasted because of all the stuff done off-shift when they hit a big fire. How they show up to the firehouse will have a huge impact on their effectiveness in doing their job. Make sure time is being taken to rest and relax. A spouse who gets this is truly a godsend for firefighters. Avoid the temptation to fill up the days off with stuff. Some is good, too much is deadly. And realize the time home is such an amazing gift for the two of you. Your firefighter will likely have some flexibility to get some of the stuff you want done, get some rest and be able to spend focused, quality time with you. It can be amazing if you do it right.

With some minor changes couples can really benefit from the 24-hour shift schedule. There will be times when the shift is a grinder and the day after will have to be a rest day. Other times, the firefighter will simply have to buck up and fight through the fatigue. If that type of schedule is what you do every day and there is no allowance for the realities of shift work, be prepared for some tough times. Work with each other on it, have each other's backs, seek some compromises that work toward making your marriage the priority. In that environment, time off is simply magical.

NOTES

1. Bobby Halton, "What We Know That Ain't So," *Fire Engineering*, July 2009, 8.
2. Anthony Evans, *The Economic Impact of Successful Commercial Fire Interventions: Phoenix Fire Department June–August 2012*, L. William Seidman Research Institute, Arizona State University, http://media2.abc15.com/html/pdf/ASU%20study.pdf.
3. Seattle Fire Department, *2016 Annual Report*, https://www.seattle.gov/Documents/Departments/Fire/About/AnnualReport2016.pdf.
4. The Council of the City of New York, *Report on the Fiscal 2017 Preliminary Budget and the Fiscal 2016 Preliminary Mayor's Management Report: Fire Department*, https://council.nyc.gov/budget/wp-content/uploads/sites/54/2016/05/057-FDNY.pdf.
5. Jessica Robertson, "One Year Later—The Oso Landslide in Washington," Science Features, United States Geological Society, March 16, 2015, https://www2.usgs.gov/blogs/features/usgs_top_story/one-year-later-the-oso-landslide-in-washington/.
6. The National Institute for Occupational Safety and Health (NIOSH), Centers for Disease Control (CDC), "Plain Language about Shiftwork," DHHS (NIOSH) publication number 97-145, July 1997, https://www.cdc.gov/niosh/docs/97-145/default.html.
7. "The 24-Hour Shift: Impact on Health and Safety, *Fire Engineering*, May 1, 2007, http://www.fireengineering.com/articles/print/volume-160/issue-5/departments/fire-service-ems/the-24-hour-shift-impact-on-health-and-safety.html.
8. NIOSH, "Plain Language."
9. Alice G. Walton, "Your Body's Internal Clock and How It Affects Your Overall Health," *The Atlantic*, March 20, 2012, https://www.theatlantic.com/health/archive/2012/03/your-bodys-internal-clock-and-how-it-affects-your-overall-health/254518/.
10. Lirong Zhu and Phyllis C. Zee, "Circadian Rhythm Sleep Disorders," *Sleep Disorders*, edited by Bradley V. Vaughn, vol. 30, issue 4, November 2012, 963–1434.
11. Ibid.
12. "Blue Light Has a Dark Side," Harvard Health Letter, Harvard Health Publishing, December 30, 2017, https://www.health.harvard.edu/staying-healthy/blue-light-has-a-dark-side.
13. NIOSH, "Plain Language," 31.
14. "The Five Most Important Melatonin Rich Foods," EU Natural, https://eunatural.com/five-important-melatonin-rich-foods/.
15. NIOSH, "Plain Language."
16. Ali Kuoppala, "Sleep and Testosterone: Each Extra Hour of Sleep Means Roughly 15% More T," Anabolic Men, https://www.anabolicmen.com/sleep-testosterone/.

17 P.H. Barnes, "The Experience of Traumatic Stress Among Urban Firefighters," *Australian Journal of Emergency Management*, 14(4), Summer 1999–2000, 60–64, http://eprints.qut.edu.au/2121/1/2121.pdf.

PART II

FIVE ESSENTIAL CONVERSATIONS FOR THE FIREFIGHTER MARRIAGE

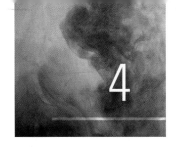

Essential Conversation #1: Re-entry Time

What is the key to any successful marriage, no matter the profession? Good communication. Communication means to speak. To be heard. To impart your thoughts to others. But the highest form of communication is something more. It is when you seek to listen, not just to be heard. To understand, not just to be understood. This is known as *conversation*, or *dialogue*. And it takes two. John Powell, author of the book *The Secret of Staying in Love*, says this: "Dialogue is an act of purest love. To live in dialogue with another is to live twice. Joys are doubled by exchange and burdens are cut in half by sharing."[1] When dialogue or conversation thrives, so does the marriage; when it breaks down, the marriage can end.

How to communicate with your spouse is a topic that could fill 20 books, but for the sake of expediency, I have narrowed it down to five essential conversations on issues that are unique to the *firefighter* marriage. I recommend that these conversations take place ahead of time, or *before* the issue becomes a fight. In the heat of the moment, it's hard to be rational or productive, hence dialogue unwittingly descends into an argument.

Essential conversation number one is re-entry time. I begin here for two reasons: one, because it involves the point at which the firefighter couple's day together begins, and two, because it is one of my most memorable first lessons learned as a new firefighter spouse back in 1990.

It was Christmastime, a season of firsts for our young family (fig. 4–1)—our first with children old enough to be aware of the magic of Santa, our first with a tree hand-cut from a farm, and our first spent with a probationary Seattle firefighter. The tree cutting had been planned and was greatly anticipated by all. The farm had been chosen, it was the beginning of a clean four-day break for Mike, and it wasn't supposed to be raining—a rarity for Seattle in winter. Perfect!

CHALLENGES OF THE FIREFIGHTER MARRIAGE

Fig. 4–1. Our young family in 1990, the year Mike became a Seattle firefighter, and the year re-entry time was born for us. (L to R) Rick, age 2; Mike; Anne; Michael, age 4.

My firefighter arrived home that morning after his 24-hour shift to be immediately pounced upon with unbridled enthusiasm. I and our two little boys (ages 4 and 2 back then) rushed at him to hurry up and get ready to go. He barely reacted, which was unusual for him. "Do we have to do this *today*?" he complained. Yes, as a matter of fact we did; no other time worked as well, which we had already established. He grudgingly acquiesced.

As the day progressed, my hopes for a magical first were dashed: my firefighter was less than enthusiastic; in fact, he was downright irritable. By day's end, as I listened to him grumbling over the complicated tree stand, my eyes filled with tears. I rushed to the bathroom, where my little boys wouldn't see, and fell to the floor weeping. My husband was not himself—after five years of marriage I felt as if I didn't know him. I had never seen him behave this way on such an important, eventful day—with repressed yet obvious anger that affected both his and my joy. I knew it was job-related, but that didn't make it any easier in the moment.

Mike had been up all night with a fire and several car wrecks. This was all new to him. On top of sleep deprivation, physical exhaustion, and exposure to

Chapter 4 — Essential Conversation #1: Re-entry Time

trauma, he had to endure the added pressure of being on probation. One wrong move and he could lose his job—the dream job he had fought so hard to get. The job he desperately needed to provide for his family. "Probies" were often mocked and harassed and forced to perform every mundane station chore without complaint. These overwhelming conditions could push anyone to the breaking point. I knew all of this, but until that tree outing, it really hadn't hit home. Mike was normally a very loving, gentle, patient husband and father who engaged wholeheartedly in everything we did. But not today. Today he was grouchy and sharp-tongued and not much fun at all. As I lay there on the bathroom floor, I wondered for the first time if our marriage would survive this new career.

That Christmas was decades ago. Those two little boys are now grown men, and we no longer go to a farm to get our trees. But we did for many years and it was wonderful. And our marriage is still going strong, in spite of the odds that were stacked against us. How did we make it? I don't know all the reasons why, but I do know that re-entry time is one of them.

Re-entry time isn't something I learned from a book, but from hands-on experience as a firefighter wife. It is a handy tool for marriage in general, but it is especially helpful to the firefighter marriage because of the unique schedule, which often requires prolonged separation, and the nature of the job itself. *Re-entry time is simply this*: allowing the returning spouse some time to enter the home without hassle or massive conversation. This can apply to either spouse of any profession, but for the returning firefighter, it is an extra special gift as it enables them to shift gears from the high-octane world of firefighting to the relatively chill mode of family life. It is a chance to decompress, a mental break from the action.

A compelling example of this need for transitional time was portrayed in the movie *American Sniper*. Navy Seal Chris Kyle goes to a local bar for a while instead of heading directly home to his wife, Taya, and their children after a long deployment in Iraq. Taya is stunned to learn that her husband is back in the states, so close, but sitting in a bar instead of coming straight to her. She is slightly puzzled, even offended by this delay, as she and the kids were excited to see him. But so intense was his experience as a soldier that he doesn't feel ready to make the necessary transition back into family life. This can be true for firefighters as well. A dear friend of ours, the legendary Kevin Shea (FDNY, retired), once gave me an example more typical of a firefighter's need for re-entry time when he said, "I used to hang out in the garage for a while before going into the house. My wife would ask me what I was doing out there for so long. I didn't really have an answer."

We are so blessed to have great protectors who act as watchful sheepdogs over us sheep. Soldiers, police, and firefighters are like the sheepdog in the following way:

"If you have no capacity for violence then you are a healthy productive citizen: a sheep. If you have a capacity for violence and no empathy for your fellow citizens, then you have defined an aggressive sociopath—a wolf. But what if you have a capacity for violence and a deep love for your fellow citizens? Then you are a sheepdog, a warrior, someone who is walking the hero's path. Someone who can walk into the heart of darkness, into the universal human phobia, and walk out unscathed."[2]

Soldiers and first responders are forced on a regular basis to go from being sheepdogs who take on wolves to being gentle enough to live with the sheep. Re-entry time helps with this drastic transition.

Anthony & Cynthia Kastros, Sacramento Metropolitan FD, California

A firefighter husband just got off work from a grueling shift. He is tired, hungry, and horny. He stinks and wants to come home to his beloved wife to begin days off. He secretly fantasizes about his lady greeting him in her teddy, with breakfast in one hand and a Bloody Mary in the other. After a shower, he wants to "get busy" and then take a nap.

In a nearby galaxy, the firefighter's wife is about to lose her mind while he's at work. The kids have been driving her nuts. She has been putting out her own fires at her job, has barely slept, has no makeup on, and this is her first day off in five days. The house is a mess and she wants to get a workout in, take a shower (by herself), and go out to lunch with him before picking up the kids.

He walks in the door. No wife in negligee. The house is a mess. And she is nowhere to be found. He finds her in the bathroom and hugs her from behind. She says, "Hey, let's go to lunch after I work out. I have to do some shopping while you pick up the kids since we're out of food. Don't forget that they have soccer practice after school."

He replies, "What? I'm tired. Can't we just hang out at home (and hump and eat and sleep, he's thinking)?" Instant conflict. Sound remotely familiar?

Perhaps your situation is two working parents. Perhaps you have a home business. Perhaps you have a bunch of kids and she is a stay-at-home mom. Whatever the case, you both have needs and wants and they can conflict. While you love each other and want the other to be happy, you are also human beings with needs and desires that may seem like they are not on your spouse's planet of consciousness. Is romance and sex wanting? Is communication a struggle? How do you get on the same page to meet each other's needs in the limited time you have together? Maybe something from the firehouse can help.

One of the most critical tools we use in the fire service is a *preplan*. Preplans allow us to become familiar with target hazards before a fire comes, thus saving valuable time and potentially preventing disaster. Critical factors such as access, utilities, water supply, construction type, occupancy, and special hazards are all identified in a preplan. This information saves time and prevents potential problems, making a much safer, more effective and efficient operation.

Marriage can benefit from preplanning as well. There are predictable—and thus preventable—circumstances in marriage that can benefit greatly from a preplan. Our marriage has certainly become more harmonious because of some of the preplans we have established along the way. Many of these have come out of necessity, due to recurring conflicts that we deemed as predictable and thus, preventable. We kept having the same fights over recurring situations and decided to preplan to prevent them. The results have been fantastic!

One primary area that needed a preplan was the transition from on duty to off duty. When Anthony would come home from work, he wanted a nap, food, and affection. Cynthia was eager to get out of the house and spend quality time together. He needed to come home to relax, while she needed to get away from home to relax. He was inundated with calls, personnel issues, and sleeplessness. She was inundated with kids, housework, and the business. This resulted in recurring conflict when Anthony got home. We had competing objectives at shift change. This soon gave way to some discussions the night before our first day off. While Anthony was at work and Cynthia was at home, we would preplan the next day, beginning with how the first few hours of our days off would go.

We discussed each other's needs, wants, and feelings about what transpired over the "tour" and what we wanted the first day. His "tour" was at the firehouse, while her "tour" was at the house. Both were wrought with hard work, stress, sleeplessness, and feeling like our days off were going to be a welcome relief. The problem was the transition at shift change. The conflict and feeling that our respective needs were not being met would ripple into the rest of our days off together. So, what should have been four days off would end up being only two quality days off since we were pissed off at each other the first two days.

Now, the night before shift change, we discuss everything: what we want, feel, need, and hope for, and what we don't want. This allows us to come up with an incident action plan for our days off, and shift change goes much smoother. For example, Anthony will say, "Honey, I am beat. I'd like to come home and nap first thing in the morning. After that, I will be good to go for the rest of our day." This gets his needs out there and gives Cynthia an opportunity to share hers too, before an action plan is established.

Cynthia will say, "Okay Babe, I'm going to go for a walk on the beach when you get home to get a break from the house, then go grocery shopping. How about we go to lunch at 11:30?"

You will notice that not *all* of our needs were met in that one exchange; however, we both got some of what we wanted, which makes us both happy and allows us to continue to preplan and communicate about meeting the rest of our needs in the hours or days to come.

Rather than just face off when Anthony walks in from work, hoping that we can read each other's minds and waiting for the other to blink, we preplan our needs and wants together. She assures him that he will "get some" while he assures her that they will have a date night without his cell phone present. This has become such a normal part of our routine conversation that it's really effortless now and the rewards have been incredible.

We have established preplans for other issues as well:

- Disciplining the kids: tactics to ensure we are on the same page and send a consistent message.
- Finances: what our goals are for tithing, retirement, vacations, remodels, etc.
- Sex: while we want to be spontaneous, factors like kids, work, fatigue, and illness all affect intimacy, so we preplan.
- Vacations: we both love knowing that we have a break coming from work where the whole family can be together.
- Individual time: it's okay to need and want alone time to recharge our individual batteries doing our own hobbies or spending time with our own friends.

Consider each of these categories "target hazards" that you know can cause predictable conflict in your marriage. Your effort in preplanning for these and many others specific to your life will pay huge dividends in a marriage that is much more harmonious and unified.

As I have written in the first three chapters, firefighting is unique among professions primarily because of the constant exposure to danger and trauma. It is a high-stress, adrenaline-packed job. When one is confronted with danger, adrenaline surges into the body (along with numerous other hormones), providing instant superhuman strength to survive. Consequently, your firefighter may come home to you in either one of two conditions: still full of adrenaline or completely exhausted and drained of all energy. Either one—amped up and agitated or drained and apathetic—can make one incapable of having a fruitful and meaningful conversation. When full of adrenaline, a firefighter may still be in full fight-or-flight mode and might actually see a brisk round of 20 questions or task demands as a threat. He may then be overly aggressive or harsh in his responses (unintentionally of course), which can result in a hurt and offended spouse. Or, conversely, a drained firefighter may be so uninterested or unresponsive to your friendly attempts at conversation or reasonable requests for tasks that the same result can occur—a hurt and offended spouse. I know this from personal experience. This is how re-entry time was born—from the conflicts that arose when different dispositions and job demands met at the front door.

My disposition, or nature, is to express pleasure at seeing my beloved with words and questions—many words and questions, and to talk freely about my emotions and problems and all that went on while he was away. His disposition, or nature, is to express his joy at seeing me with a hug and a kiss and very few words. It is hard to suppress our natures, especially when fatigued. According to Dr. Luan Brizendine in her book, *The Female Brain*, women speak more than men. The study went on to reveal that a woman's brain is wired to talk more, and that women experience a rush of feel-good chemicals when they speak. Dr. Brizendine emphatically states, "We're not talking about a small amount of pleasure. This is huge. It's a major dopamine and oxytocin rush, which is the biggest, fattest neurological reward you can get outside of an orgasm."[3] The research done on male and female brains reveals obvious physiological variances in the areas devoted to speech and the processing of emotion. Another study reports, "What we observe when we analyze various couples is that women like verbal communication more than men... It takes energy for men to say how they feel, share their problems, and even to share good news. Many think talking is synonymous with putting themselves at risk. It's an area they feel particularly unskilled in, and they're afraid of slipping up."[4] Vive la difference! I think anyone who has been married for more than a year can tell you of these inherent differences without being a brain surgeon or researcher. But as with every assertion there are always exceptions, and feel free to adapt your particular situation and personalities. The key is to recognize the conflict that occurs on re-entry and give your spouse what they need, whoever it is that talks a lot or doesn't.

As a young wife, when my firefighter would return after 24 hours (or more), I would show my delight at seeing him by greeting him with a recounting of all that had happened in his absence, good or bad. He felt no desire to reciprocate the information and I used to get offended by this. Instead, he was usually quiet and a little distant or even irritable. I assumed this meant he was not as glad to see me as I was to see him, so many fights ensued. But over time, I came to understand that this was not the case, and I adapted by giving him what he needed most from me after a long shift—not conversation, but re-entry time.

Sleep deprivation, backlash, and trauma sometimes combine to make my firefighter rather tired and often completely drained. It was and is to this day a gift for both of us to just let him be for a while. A positive tone is then set for the entire day, one of graciousness and appreciation. We've avoided many fights, and, much to my delight, he is so grateful for the time to shift gears that later in the day he reciprocates by giving me what *I* need—long conversations and genuine attentiveness. It also helps if he calls me from the station at least once the day before. I just want to know he's okay. This, too, used to be harder to do before the days of cell phones. But now that we have them, he can call me, no excuses. He does so faithfully, as he knows what it means to me. Having heard from him, I'm less likely to pounce the minute he walks in the door.

We don't always get it right—sometimes I still fail to give him his space and sometimes he fails to give me enough attention—but that's the nature of marriage. Give and take, understanding and compromise. When I can, I take the hit and when he can, he does. In fact, more often than not, Mike has had to forgo re-entry time and hit the ground running when he gets home. And this he has learned to do, cheerfully, despite his fatigue. Sometimes re-entry time is possible and sometimes it's not, and here are some of the reasons why.

Little children can't wait

I'll be honest: re-entry time gets easier with age—the age of your children, that is. When our kids were little, their needs were immediate. With two little boys I was hopping around nonstop, 24/7. I, too, was exhausted and looked forward to a break when my husband came home. I often thrust them into his arms the minute he walked through the door. Sometimes Mike handled this very well, sometimes not so much—we weren't and still aren't perfect—but he tried to take them off my hands as much as he could because that's what parenting is all about. They loved their daddy and wanted to hang on him like a human jungle gym; re-entry time meant nothing to them!

Plans can't always be changed

Holidays, my work, school events, doctor appointments, house repairs—you name it, things did and still do come up. Life goes on. It can be just as frustrating for me to have to adapt my plans as it is for him. He has had to take the hit many times, in spite of a rough night, and this he has done, graciously—especially since his probation ended so long ago!

I've had many sleepless nights too

Sleeping alone has posed many scary threats for me over the years. I once had a stalker, which resulted in night terrors; I once heard footsteps in the loud, crunchy gravel outside my window and had to spend the night with relatives. Someone once followed me home, which again resulted in my having to go to a relative's house. I've weathered snow storms, power outages, an earthquake, busted pipes, and even a flooding basement from a rainstorm—all on my own, and always, it seems, in the middle of the night. I've been up all hours with sleepless babies, sick or frightened children, and even with the numerous demanding puppies we've had over the years. And I have, on many occasions, spent anxious nights fearing for my husband's life, alone in the dark. Mike's re-entry-time naps have often been relinquished for the sake of my sanity, as I too have needed to catch up on sleep.

It takes two to tango, two people willing to put the needs of their spouse ahead of their own. Spousal support is vital in the high-stress world of firefighting. Your spouse is essential to your success, so don't take them for granted. They patiently put up with your irritability and fatigue, and you must remember to do the same for them too. Being apart for so long requires an extra effort and lots of skill to reconnect gracefully.

Part of the dance of marriage is learning to give and take, to meet each other's needs, though they may be contrary to your own. Mike needs peace and quiet when he first arrives home; the last thing he wants when he's wiped out from a 24- or 48-hour shift is a long conversation. He needs time to rest, relax, and recover. I give this to him, gladly. Then later, when he's rested, he gives me what I need—conversation—and he gives me this gladly. But my willingness to wait awhile to get what I want has made all the difference. He is so grateful for re-entry time that he lavishes me with long phone calls from work the night before and details, details, details about his world, after a bit of time off.

If his shift was particularly rough, plans may need to be changed, and they are, if it is possible. This is what my firefighter often needs from me and it is a gift that I have learned to give him. It has saved many family outings as they were rescheduled to a time when he was better suited to enjoy them.

A good marriage requires effort, planning, and engineering, especially in the arts of communication where men and women's differences are the most prevalent. Renowned philosopher G. K. Chesterton once said, "...the greatest feat of engineering in human history is the bridge that has been built between man and woman."⁵ Build your bridge, don't burn it. Keep your love alive by knowing your differences and working with them, not against them. It is so very worth it to get to the other side. Having that conversation about re-entry time is a great place to start.

HONEY, I'M HOME!

Of all the articles Anne has written, the one on re-entry time seems to have resonated most with couples from all walks of life. It seems a universal concept and one that will have dramatic impact if given the attention it deserves.

As you'd imagine, it is difficult to read about what a jerk I was during something as sacred and special as getting the Christmas tree with my little kids. Obviously not my best moment and one, like many others, I wish I could take back. But we don't get them back. We don't get to do those bad days over and that's all the more reason to work to minimize them. And, of course, to learn from the things that got us to the bad place to begin with. I believe re-entry time is a big help in keeping many potential bad days from going completely off the rails.

YOU TALKIN' TO ME?

It may surprise you to learn that I do not recall the "Christmas tree debacle" as being as bad as Anne recalls it. I remember being very tired and trying to schedule it for another day. My attitude wasn't the best but I felt I gave it a good try. And this is the point you should really take to heart: you are not always aware of how you're coming across to the other people in your life, especially when tired or impacted by what has happened at work. My behavior was obviously serious enough to ruin the day for Anne, but it didn't seem that bad to me.

It is this gap that re-entry time really impacts. I can't put into words how beneficial it is to have a bit of time to acclimate to the very different environment of home versus work. It provides a buffer zone of moving from giving orders and making rapid decisions to being part of a marital and parenting team and navigating domestic life. They are such different animals that shifting in your mind is important. Especially if your shift involved difficult runs.

Many of our friends have worked to make re-entry time apply to their situations. There are so many variables to how households are run that we could never hit them all, but here are a few ideas:

Find a quiet place prior to entering the house. Your shop, garage, backyard, a nearby park, a store—there are endless possibilities.

Use your commute home to purposely shift your mind. Whether it's a specific type of music, radio program, spiritual material, or just silence, get your mind in the process of being home.

Find a small activity that helps you reset. This can be with pets, hobbies, chores, or whatever helps you most. We know a firefighter who goes to the barn to hug his horse (fig. 4–2) before going into the house to hug his wife. He says it calms him like nothing else. It may seem unusual, but this animal is a rescue horse that had a significant impact on the health of their family. Now his actions make a little more sense.

Fig. 4–2. Re-entry time in action: firefighter Stephen Funk hugs his horse to decompress.

Do a quick workout to burn off some steam, stress, and emotion. It can be done at the firehouse, gym, or wherever is best for you.

Of course, there are going to be times when it may not be possible to have that break. It is particularly difficult when both parents work and a hand-off of kids or other responsibilities needs to occur. The point here is that this transition, and the dangers it can pose to marital harmony, is very real. You can ignore it and rationalize that it's just a part of modern busy life, or figure out a way to adjust your schedule and give each other the necessary gap to make things work. Many couples, who are just as busy as you, have figured it out and felt the immediate benefit.

IF IT'S GOOD FOR THE GOOSE...

A firefighter who is blessed with a spouse who allows them re-entry time should be enthusiastically willing to do two things in return:

Get re-acclimated to home as quickly as possible and then engage in ways that are desired by your spouse. Whether it's talking about the shift you just had or planning the day ahead, or simply giving them a much-needed break, it's your turn to give.

Re-entry time goes both ways. Your spouse can use a similar courtesy when they return from their job or have been with the kids all day. Everything we discuss should be mutually beneficial and reciprocal.

It seems like such a small thing to give just a bit of a break to your mate, but it's as big as anything we talk about in this book. Once Anne and I got this figured out, we avoided a lot of early morning hassles—the kind that blow up and ruin what would have otherwise been perfectly good days.

NOTES

1. John Joseph Powell, *The Secret of Staying in Love: Loving Relationships through Communication* (Niles, IL: Argus Communications, 1974).
2. Dave Grossman and Loren W. Christensen, *On Combat: The Psychology and Physiology of Deadly Conflict in War and in Peace*, 3rd ed. (Warrior Science Publications, 2008).
3. Louann Brizendine, *The Female Brain* (New York: Harmony Books, 2007), 64.
4. "Women Want to Talk, Men Want to Run Away," Exploring Your Mind, December 24, 2016, https://exploringyourmind.com/women-want-talk-men-want-run-away/.
5. G. K. Chesterton, *The Superstition of Divorce* (London: William Clowes and Sons, 1920), 24.

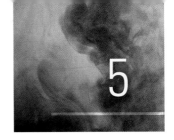

Essential Conversation #2: Harshness and Gallows Humor

Firefighters are a tough breed. They're usually Type-A go-getters, hard-working, aggressive, confident achievers. They tend to be athletic and very active. They are drawn to the fire service for these very reasons, as firefighting requires aggression, physical prowess, and teamwork. It is an exciting, adrenaline-packed job that is both exhilarating and fulfilling. Firefighters can pour themselves into this demanding job and feel rewarded from all the self-sacrificing exertion they experience every single day. For these types, it is a dream job: never dull, ever changing, and always challenging.

In this profession, where life itself can be on the line, firefighters form bonds with each other that few can truly understand. It is indeed teamwork at its best. Firefighting is a rough business that requires incredible amounts of strength, courage, and let's face it, some pretty thick skin. When exposed to danger, as we all know, adrenaline supplies an extra amount of aggression needed for courage and strength. That aggression is essential when handling heavy equipment, kicking down doors, and having to work with other firefighters under duress. In the midst of danger or intense aid runs, decisions must be made quickly. Orders are given, even shouted, to be heard over the din of chaos. No one is offended by terse, abrupt modes of conversation; words are kept to a minimum when life is on the line. Each member of the team rises to the commands and responds immediately, taking appropriate action. This type of communication is necessary, even somewhat beautiful in its strength and blunt simplicity in getting the job done. Does barking orders to a fire crew work? Absolutely. Does barking orders to a spouse? *Absolutely not.*

Trauma has its own repercussions as well and let's face it—our firefighters witness some pretty gruesome sights. They have near-misses with death and are present many times when it doesn't miss. To cope, to survive, it's only natural that they get a little hard-hearted. Some of this hardness exhibits itself in anger and some in humor. Anger is easier to work with than sorrow. So is humor.

In fact, humor is the healthiest way to cope with horror as it is not only harmless but is actually healing. Laughter is the best medicine for a wounded heart. As a result, firefighters develop a certain dark humor in the midst of trauma which is known as *gallows humor*.

Gallows humor is a phrase coined from the actual device itself: when victims were to be hanged on the gallows, they often shouted out bits of humor to cheer their distraught friends and family members who were in attendance. For example, William Palmer is reported to have looked at the trap door of the gallows and said, "Are you sure it's safe?" Gallows humor, for the firefighter, arises from stressful, traumatic, or life-threatening situations. Humor is a coping mechanism used to alleviate some of those stresses. It is far less painful to laugh than it is to let yourself feel the overwhelming emotions of the tragedy itself.

The humor may be too edgy or offensive to share with outsiders, but between firefighters, it becomes an inside joke. It may seem inappropriate, but this humor is indeed quite natural; it makes disagreeable tasks more agreeable and is present in many professions. It keeps firefighters sane by filling the void of needed expression safely and harmlessly. It does not, however, translate well to the outside world. Gallows humor can make one appear cold, uncaring, insensitive, and even disrespectful. It is best kept between firefighters, on the job, with those who share the joke. (Funny addendum to illustrate this point: I asked Mike to give me an example of gallows humor from his own experience; he could not think of *one* that would be appropriate for the public to read. Neither could any of his crew! All their examples, they felt, were way too dark.) Finding a bit of humor lightens the mood and helps the crew carry on. Laughing at a tragic scene may seem inappropriate to the outsider, even twisted. Harsh jesting works in the firehouse. Outside the firehouse, it is simply harsh.

Aggression from adrenaline and anger or dark humor from trauma, combined with a third ingredient—sleep deprivation—are almost guaranteed to create harshness. Everyone gets grouchy when tired; this is normal. In fact, all of these behaviors are quite ordinary when dealing with extraordinary circumstances. Firefighter couples need to be aware of these realities and have coping strategies in place in order to protect intimacy which is so easily shattered by coarseness.

Danger and trauma exposure is ongoing for firefighters; they are not just one-time events. Firefighters adapt or survive by developing these behaviors, getting a little hard-hearted along the way. To be jaded, even a little detached from their own feelings, is part of the job. At the firehouse, these behaviors don't mean anything. No one gets offended because they all do it. The work environment actually creates these responses, and they are natural. What is unnatural is when the harshness comes home. In the home, these behaviors take on a new light. The spouse, who is unaccustomed to danger or trauma, can see these types of behaviors as offensive, even hurtful. The spouse can think, "It's me. He's unhappy with me. I must have done something to provoke this harshness." The spouse takes it personally and responds by pouting or withdrawing. Or the spouse might think, "It's him. He's being a jerk," and decide to retaliate with the same type of harshness, sarcasm, or anger. I know this, because I've done it.

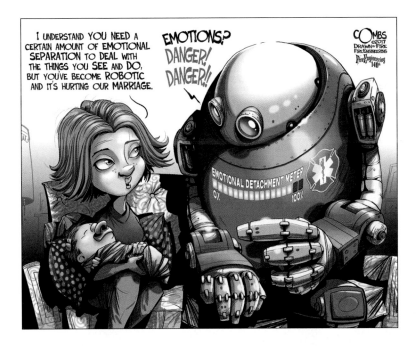

Neither withdrawing nor retaliating do anything to improve the situation; these responses will only push you farther apart. So what do you do? Harshness should not and must not be allowed to continue, as it is a very destructive force within a marriage. Have this essential conversation: spouses of firefighters must tell them when they are being too harsh, in a constructive way.

To begin this essential conversation, always approach your firefighter by assuming the best. Try to understand that this harshness is a defense mechanism, not a character flaw. And remember, it's not about you. Try very hard not to take it too personally, while at the same time working to make sure that this doesn't impact your relationship. Again, it is natural to be hard-hearted at work, but it is unhealthy to be so at home.

As stated before, it is best to have these conversations when your firefighter is well-rested and has had a chance to fully recover from the draining effects of adrenaline. Then they are better disposed to listening and understanding what you are saying. Re-entry time helps temper this harshness. Use it as a tool to help your firefighter shift gears from the tough world of firefighting to the soft place of home (fig. 5–2). It is the first step in combating the distance that can form between the sheep dog and the sheep as I've told Mike many times, I am indeed a sheep and he needs to be gentle with me:

"The sheepdog. He looks a lot like the wolf. He has fangs and the capacity for violence. The difference, though, is that the sheepdog must not, cannot and will not ever harm the sheep. Any sheepdog who intentionally harms the lowliest little lamb will be punished and removed. The world cannot work any other way, at least not in a representative democracy or a republic such as ours."[1]

Fig. 5–2. Mike as training captain looking harsh with recruits. This works with them, but if he looks this way at me? Not so much. Courtesy of Lt. Tim Dungan, Seattle FD.

Have an agreement ahead of time to use key phrases that you both recognize as indicators that the harshness is creeping back into the relationship. It is sometimes difficult to create these phrases in the heat of the moment; thus, planning and preparing them is more effective. For example, my husband Mike and I have agreed on my saying, "Hey, I'm not a firefighter." (Our dear friend, Captain Mike Dugan of FDNY, told us his daughters say to him, "Dad, stop using your fireman voice.") We've discussed this issue at length over the years, and these types of friendly little reminders are a way of gently broaching the subject without getting into a major argument. He knows and I know what these phrases mean: you're starting to offend me with your coarseness, so please, lighten up.

Mike has specifically asked me to tell him when he's being a jerk; he needs to hear it from me. Otherwise, he can be unaware of the fact that he is indeed starting to treat me like a firefighter at a fire scene, barking orders and such. I've been honest with him and told him I don't like it. I'm a tender-hearted person and I need to be treated like one. He likes me that way; it's one of the primary reasons he married me. Words are very powerful, especially to me and especially *his* words. He must choose them very carefully if I am to continue to trust him with my heart.

Our relationship has weathered these storms because we are willing to have this conversation, though it can be difficult. Mike has come to understand that his behavior can be affected by his work and that that behavior may be harmful to me emotionally if it continues unchecked. I've come to understand that it's often difficult to shift gears from the work atmosphere to home life, especially when you add sleep deprivation to the equation. He does the best he can and never *intends* to be harsh with me. And because he has taken the time to explain his stories, I too have developed a healthy understanding of gallows humor. I laugh right along with his anecdotes, though they are macabre and should not be told at baby showers or ladies' tea parties or family gatherings. I've accidently done so before and witnessed horror instead of laughter. I guess my skin has thickened a bit too over the years. Let your spouse in on the jokes, firefighter, and they will come to share them with you instead of being offended by them.

Intimacy is a very fragile, silken thread that binds two souls together. It must be protected to survive. Unchecked, undealt-with harshness can damage this thread, even threaten to break it. Firefighters need a soft place to land after the hardness of their work. It keeps them from becoming too inured. Home should be that place. It takes a concerted effort on both your parts to maintain that softness; otherwise, the harshness of the job will creep in. Emotional detachment is contagious, so if one of you checks out, the other eventually will too. And emotion is required to *feel* intimate.

The word "intimacy" is derived from the Latin *intimus*, meaning "inmost." Inmost refers to the core of a human being—his or her soul. The soul, that place where the heart, the mind, and the will rise up to make an individual capable of great love and great closeness and connection with another. Intimacy is to share this soul in private, tender, heartfelt ways with another. It is an emotional closeness where two people are constantly alert and responsive to the feelings, needs, and well-being of another. They are understood and seek to understand. We believe human intimacy renders the commonplace relationship unique and extraordinary and nearly indestructible, for all eternity. It is central to a strong marriage.

Dave & Gina McGrail, Denver FD, Colorado

DAVE:

Let me begin by saying that I am very fortunate to have found, fallen in love with, and married my wife. She is a very loving, kind, and compassionate person. She demonstrates this in her everyday actions with her family and all other people she encounters.

Our marriage has a very strong foundation based on our mutual belief in God, as well as our overall spiritual and religious beliefs. We also have very likeminded philosophies with regard to money and politics. Our similar beliefs regarding these three very important items (religion, politics, and money) are the primary elements that I believe help us have a strong marriage and relationship. More often than not, we will be thinking and saying the same thing, simultaneously. With that said, I also believe that my wife has, over the years, given more to our relationship, and has put up with occasional behavior from me that was unacceptable.

Let me explain. As a young man, I first fell in love with the desire to become a fireman and follow in my dad's footsteps. I was one of the lucky ones who got hired and accomplished the goal of getting "on the job." I was truly in love with the fire service, the fire department, and being a fireman. I loved the job. With about fifteen years on the job, I met and married my wife. Unfortunately, because of my lack of maturity, I still put the job first. After a few years of marriage, we were blessed with children. We have a beautiful daughter and a very strong and brave son.

My completely absurd prioritization manifested itself in a very ridiculous way several years ago. I had established a goal many years ago that I would never miss a work shift, just like my dad. He worked for 42 years and never missed a shift, never called in sick, not once. Who calls in sick? You might miss something, like a fire.

Keep in mind, in addition to being the mother of two (or three, some might argue), my wife was a working professional her entire adult life. She had a demanding full-time executive-level job. So, on this particular morning, she had to travel for business. We had previously setup childcare for our kids, but the person staying with them got sick at the last minute. My wife had to be at the airport in a few hours. But that didn't matter because I had to be at the firehouse, and what could possibly be more important than that? I literally abandoned her that morning, left her to take care of the kids, cancel her flight, change her schedule, etc., etc., etc. Yes, what a *dope*! Not my finest moment. She should have left me that day. She shouldn't have forgiven me. But she didn't leave and she forgave me almost immediately.

Moments like these finally changed me in ways that should have happened long ago. My family will always be with me, the job will not. Love your spouse and family like there's no tomorrow. Call in sick when you or your love one's are sick or there's a family emergency. Take every second of vacation time you have coming to you, and spend that time with your family. Those are benefits of the job and ultimately, the job, as great as it is, is still just a job. It will be there when you get back from vacation. And when you get back, put your professional firefighter cap back on, and give it 100%. When you're with your spouse and family, give them 100% too.

The moral of this story is this, *family first*! Period. End of story.

GINA:

When I married a firefighter, I went into the marriage knowing that the job is physically demanding and dangerous. There are certainly risks, but I believe, hope, and pray that the risks are minimized with proper training and the use of sound practices. Of course, the risks are never eliminated and accidents still happen. We have firsthand experience with this. It's difficult. In these times, I pray for God's protection over the crew, the incident, the family, and ask for the strength to keep going.

The peril I wasn't entirely naive to but also not completely prepared for is the degree to which departmental politics can put a strain on your spouse and your marriage. When someone hurts your spouse, they hurt you twice. It's stressful. No amount of training can mitigate this risk. As we know, some "leaders" (not all) would rather do what is politically correct or advantageous for their career or that of their buddies, rather than what's right, fair, ethical, or simply kind. Despite this, I still believe that it is the best practice to live by the Golden Rule and strive to do so, even though it's tempting to try to get even. In the end, you will sleep better and your marriage and family will be better for it. Also, it's important to keep in mind that the job is just one aspect of your life; it's not your whole life.

Finally, and perhaps most importantly, make time for your faith, family, fun, fitness and friends! And stay connected with friends both inside and outside of the fire service. Remember, balance is healthy!

Dr. Helen Kaplan, author of *Disorders of Sexual Drives*, writes this on intimacy:

It is an important ingredient in the quality of love and of life. A high degree of intimacy between two lovers or spouses contributes to the happiness and emotional stability of both. All activities are more enjoyable, and life is richer and more colorful when shared with an intimate partner....

An intimate relationship acts as a buffer, providing shelter from the pressures and tensions of daily life. Without intimate relationships, we tend to get lonely and become depressed. The availability of intimate relationships is an important determinant of how well we master life's crises.[2]

Marriage is a magical, rewarding, and beautiful relationship. It is unique among all others, and intimacy is what makes it thrive. Without it, two people simply become polite but distant strangers. Conversation—healthy conversation—relieves pressure when tension mounts. It is required to maintain that silken thread between you. You must be willing to have the essential but difficult conversations so you can continue to be true soul mates, not just roommates. Harshness from a partner interferes with your ability to honestly communicate your soul. It is the absolute enemy of intimacy. Have the harshness conversation, decide what works best for you, and use that method to keep harshness from tearing you apart.

HUMOR AT THE HANGING

It is true that Anne asked me to give her some examples of the gallows humor we used at the firehouse over the years. That seemed like the easiest thing in the world, as there are hundreds of examples that have occurred throughout the years. As I began to process each one and consider how they would look being written down and shared with the world, it was clear none of them could be used. They were simply not appropriate for others to hear, at least not without an exhaustive explanation of context. That's really the key point to keep in mind about this type of coping mechanism and its close ally, harshness: They are appropriate among the team and much less so outside of the team.

THE JOKE'S ON YOU

When I was assigned to the training division, I knew there would be countless things I'd miss about leaving operations and responding to fires. What surprised me most was that one of the biggest losses was riding back from serious runs. Those times in the rig, with the team you just went into battle with, end up being some of the best moments on the job. It is full of laughter, joking, smack talk, smack down and some seriously rough humor. I missed it as much, if not more, than anything else from riding the rigs. It was therapy in a rough-hewn, visceral sort of way.

I make no apologies for the stuff we say in the rigs and in the firehouse to try and bring some relief to what can be an absolutely horrifying job. No apologies whatsoever. We see stuff that no human being should ever have to see. We deal with people in situations that are heartbreaking, terrifying, and brutally demoralizing. Any coping mechanisms that can alleviate some of the stress from these things is not only appropriate, but necessary for survival.

And this type of humor is present in all types of jobs and work environments. You'd likely be shocked if you got in a room with school teachers and heard some of the stuff they say about your precious, unique, one-of-a-kind snowflake son or daughter. Or how the nice person selling you shoes or suits deals with the day-in, day-out drudgery of their job. The jokes would likely be offensive if you overheard them, because they are not intended for you and you don't have the context that they do. It's very much the same for firefighters. And we have to be aware of this because it can be way too much to bring home, especially in light of how dark our stories can be.

One of the ways this comes out is our penchant for roasting each other over things we say and do, or personality quirks, and so on. You can be the target of some hilarious ribbing in an instant and then be all over the next guy before the laughter has died down. It is a form of camaraderie that can come out as cruel or mean if you are not a part of the team or don't understand the culture. We'd literally die for each other. Not talk about dying, but actually die. That intensity of commitment also plays out in different ways than might be normal in an office or at school. We enjoy getting a rise out of each other. I have slipped on occasion and tried this mockery on Anne. It doesn't fly with her—not one little bit.

Again, all relationships are different and some can stomach much harder stuff than others. What we've found important is to find the boundaries and do our best to keep the dark humor out of the home. That requires us to be honest with each other, to have the confidence to say that what is being said is bothersome without fear of a hostile reprisal.

TOO MUCH HOT IN THE SAUCE

As mentioned above, the second part of this is harshness. It is a typical condition that comes as we harden ourselves against the many difficult situations we encounter doing our work. That protective shield is also normal and may require an extra measure of understanding from your spouse. One of the very best things you can give your firefighter is an atmosphere of empathy and a willingness to give the benefit of the doubt. Allow them to work through some pain, frustration, or fatigue without taking it personally. There is nothing quite like the sympathy of a loved one when the dark clouds set in. Conversely, there is nothing so aggravating as having your mood darkened further by a home where there is no understanding of the pressures you face as a firefighter. This is in no way an endorsement of or excuse for abuse. Abuse should not be tolerated and if you are being abused you should leave. Get to safety first and figure out the rest once you've done so.

Firefighters, this is the wakeup call that harshness is a problem for us. It comes on like a protective cloak at work and ends up being a battering ram when we return to our normal life. It is your job to do what it takes to lessen the impact. Find ways to gauge how you are reacting and interacting with your family. Seek other coping methods that don't involve a hard shell or sharp tongue. Be ready to apologize when you've brought a hardened attitude home, and give your spouse the benefit of the doubt, as they don't always understand what you've been through. How could they?

Both gallows humor and harshness can be used to cope with stress. In the right environment, they can alleviate some of the heavy emotions that come with seeing people in very bad situations, day in and day out. But home is not work and your family are not patients or victims. Actively seek ways to identify when you're off track and work with your spouse to create the atmosphere that suits your marriage, one of trust and openness and intimacy.

NOTES

1. Grossman and Christensen, *On Combat*.
2. Helen Singer Kaplan, *Disorders of Sexual Desire and Other New Concepts and Techniques in Sex Therapy* (New York: Simon and Schuster, 1979), 183.

Essential Conversation #3: Handling the Tough Runs

It happened after 24 years as a Seattle firefighter. I could hear it in his voice that night when he called me from the station; he was quite animated, even loud. Usually on the phone he is calm and soft-spoken as he tells me that he'd had a few runs that day but they were "no big deal." But not that night. That night he was agitated, full of adrenaline, and wide awake, not his usual tired-sounding self. I knew it before he even spoke the words: he'd had a tough run that day, one of his toughest ever.

"It was pretty bad," Mike said to me on the phone, which is saying a lot, as he always tones everything way down. When Mike or any other firefighter says that a run was "pretty bad," that can be interpreted into civilian terms as "horrific beyond description, the stuff of nightmares, a scene at which human tragedy and devastation occurred." Keep this in mind, as your firefighter does witness these types of events all the time, as was outlined in Chapter 2. When they tell you it's bad, believe me, it's *really* bad. Though they might not clearly say it or even realize it themselves, they have, indeed, been traumatized.

"What happened?" I asked.

"Car fire." He answered. "A young couple was inside. They were rear-ended by a guy who's done it before, a block from where Michael and Samantha [our son and daughter-in-law] live."

"Is the couple all right?" I dared to ask, though I could tell from his voice that they weren't.

"No," he said, "they burned to death."

Being burned alive is every firefighter's worst nightmare. The only thing worse would be to have it happen to one of his children. Firefighters typically report that runs involving children or harm coming to their fellow firefighters are the most traumatic of all. And as we parents get older, so do our children, so for us, a "child" can be anyone from a young adult on down. The sphere of traumatizing runs only increases with time in the fire department; it does not lessen. Mike had to face two horrors on that run: the horror of these "children's" fiery deaths, and the horror that it could, in fact, be *our* children in that unrecognizable car. This possibility hit him like a truck. So intense was his fear that his hands were shaking as he tried to call our son. His heart nearly burst as he waited for Michael to answer or *not* answer. Thankfully, our son did answer, which kept his dad from nearly dropping out. Mike's relief was huge, but the emotion was all still there. Those young people were somebody else's children, but in those moments he glimpsed on a purely personal level the weight of their loss. For that reason, this run had a major impact on him. He responded to this particular run with an extra measure of fear, helplessness, and horror, which are the components of potential PTSD (fig. 6–1).

Fig. 6–1. Mike at a very tough run that still impacts him today. Courtesy of John Odegard.

His crew spent the rest of the evening in a "defusing" meeting—a mandatory protocol after a particularly traumatic event. This helps immensely, as it gives firefighters a chance to blow off some steam within a supportive structure. But the lingering emotions still came home with Mike the next day, as my husband most certainly has a tender heart, and it was wounded.

Mike sees horrible stuff all the time, but I rarely hear of it because not all of it affects him as deeply as this one did. This time, however, I knew. How did I know? Because he *told* me. Over the 30-plus years of our marriage we've developed a healthy mode of communication for handling the tough runs. It took us awhile to iron out the kinks of dealing with trauma, but we've got it pretty well down pat.

Step one: Firefighters must tell their spouse when they've had a particularly upsetting shift.

If they do not, the spouse may and usually will misinterpret their mood upon arrival at home. Spouses, though close, are not mind-readers. We can, however, pick up on negative moods and behaviors and often mistake them for something personal. The typical firefighter desires to protect, to save, and to rescue, so it is only natural for them to wish to keep the ugly truths of death and suffering from their mate. It is born of a noble concept: chivalry. In the past, Mike deliberately kept his tough runs from me; he believed that to share them might actually do me harm. This was a good intention, but a misguided one, as misunderstandings can and often did result. I would rather know what's really bothering him than not know.

Exposure to trauma and death causes emotional reactions; this is totally normal. The only time one does not respond with sorrow to human suffering and tragedy is if one has no heart. Most firefighters are *all* heart—that's what led them to serve humanity in the first place. But caring comes with a price, and that price is pain. Typical reactions to trauma include anxiety, irritability or anger, moodiness, sadness, and depression. One can become distracted, withdrawn, and have trouble making even simple decisions, like choosing what you want for dinner (dinner choice is a common problem in our house). Sleeplessness and fatigue may also be a result of trauma as grief is exhausting and sleep does not come easily to a troubled mind. All of these behaviors have a *direct* impact on family life, and, if ignored, the effects can be long term and quite damaging.

It is not uncommon for firefighters (and cops) to suppress emotional pain by turning to alcohol or drugs (or both) for comfort. This is an unhealthy outlet for stress and can lead to divorce, job loss, health issues, and even death. The much better option is to simply tell your spouse what's going on with you—talk about it, let it out, for it is no secret to them that something's wrong anyway.

Tom Kenney, Providence Fire Department (RI), writes that "once you begin to share your real feelings and fears with your spouse, it relieves you of the burden of always having to seem undaunted by the traumatic things we are exposed to from time to time. It allows you to strip down that wall of invulnerability at home while still maintaining control at work."[1] Your spouse can be your best source of support, but only if you allow it. We've heard many people express, in different ways but to the same conclusion, that human beings can endure incredible amounts of stress if they have just one person in their life who truly sees and understands and supports them in what they're going through—*just one*. If you let your spouse be that one, not only will your marriage and family life be better, but so will your job performance. Within the nurturing oasis of home you can recharge your batteries, restoring the strength needed to get up and do it again. But the first step begins with you, firefighter, in opening up to your mate. Be open and honest, share your true feelings. You don't have to go into gruesome detail; all your spouse really needs to know is how and why it impacted you emotionally.

Step two: This one's on you, firefighter spouse.

They've dared to tell you they've had a tough run—they've trusted you with this sensitive information. So what do you do about it? Communication is, after all, a two-way street, and it is not always the firefighter's fault when it goes awry. Perhaps they've attempted to share their painful feelings with you in the past and you didn't handle it very well, so now they're reluctant to do so again.

Over the years I've witnessed the weight of stress and sorrow that has been loaded onto my firefighter husband's shoulders. They are strong shoulders, stronger than most, but even he occasionally nearly breaks from the pressure. On these really heavy days he needs a little help, as does any firefighter, and we spouses can either help take the load off or add to it. So what does actually help my firefighter and what does not?

It has been my experience that being truly helpful is having an attitude of *selflessness* or *self-sacrifice*. This flies in the face of the modern-day gurus who chant, "Put your needs first, always," but Mike and I believe the opposite is true for a marriage, any marriage, to be successful. To us, love is being there for each other even when it's inconvenient.

Who doesn't know the lyrics to Simon and Garfunkel's "Bridge over Troubled Water"? These are some of the most beautiful words ever put to music and I believe they should be the theme song for every firefighter marriage when it comes to handling the tough runs: (And no, this song is not about drugs as is commonly believed; 'Silver Girl' was the nickname of an actual friend of theirs.)

When you're weary, feeling small
When tears are in your eyes, I will dry them all.
I'm on your side when times get rough
And friends just can't be found
Like a bridge over troubled water
I will lay me down

....

Chapter 6

Essential Conversation #3: Handling the Tough Runs

When you're down and out, when darkness comes
And pain is all around
...If you need a friend
I'm sailing right behind
Like a bridge over troubled water
I will ease your mind.

These words speak of noble friendship, of caring, of committing to be there for someone when they truly need you. We all agree that these are fine sentiments and we may tear up at the thought of someone saying these words to us. But do we really follow through when called upon to do so? What does it mean to lay yourself down, to be a bridge over troubled water for someone you love? For me, it means to truly be a source of comfort to my firefighter husband when he's hurting. I try to be the bridge that takes him from the side of pain to the side of healing, thus helping him get over the tough runs. My idea of comfort is different than his, so I've carefully learned over the years what he wants, not what I *think* he wants. Discover what makes your firefighter feel better, then make an effort to do just that.

Rob & Jodie Fisher, Snohomish County FD #7, Washington

Interesting thing, stress and marriage. It can make you or it can break you.

When we were engaged, we went through premarital counseling. Not because we thought it was needed, but because our minister, who was also a fire department chaplain, said to trust him, and we did.

During the counseling, one of the things he said to me was "Firefighters are always on alert, always."

I smiled and said, "I understand. You mean at the station."

He smiled back and said, "No...*always*." Think the movie *The Sandlot*: F O R E V E R.

How do you as a partner deal with someone who is always on? When do you cry uncle and tap out? How do you know when your mothering instincts and saying it's going to be OK don't cut it any longer? Firefighter stress doesn't just mean a rough call. It can be discontent with coworkers, frustration over a job assignment, someone not cleaning up the dishes or checking the rig, or a coworker dealing with external issues. Do you try to make the situation better or do you call in the cavalry?

This firefighter wife learned valuable lessons twice recently (always learning!) The first lesson was during the mudslide incident in Oso (WA) in March 2014. Rob was deployed on the tech rescue team on Sunday, March 23, 2014. A quick kiss goodbye, a hastily packed bag of clean socks, underwear, and toothbrush, an "I'll call you when I can," and he was off. I quickly realized, while watching the news, I wouldn't see him for some time. We live in downtown Seattle and the slide was approximately 60 miles from our home. Rob and his crew made the Berg Homestead, in Arlington (approximately 15 mi from Oso) home base. This home base happened to be the home of his driver, whom Rob has worked with for 19 years on L72, and who was also deployed on the slide. The job was long hours in rain, mud, unbearable conditions, and heartbreaking work. Rob spent 21 days working the slide. During that time, he and his driver would make their way back to the homestead in the dark. They would clean their gear, do their paperwork, eat a meal, and *talk*. They talked about the people, the work, the site, what had been done, what needed to be done, what they had seen, what they had accomplished, the victories and the defeats.

Did I want Rob home during this time? Absolutely. I wanted to make it better and be his sounding board. Could I have done that? Absolutely not. I realized, for this moment in time, the best thing was the brother/sisterhood. A partner's ear couldn't compare.

The second time I realized the cavalry wasn't me was in the spring of 2016. The patriarch of our family, his father, unexpectedly passed away, and Rob was passed over for a promotion he'd put immense effort into. Both happened within weeks and Rob plummeted into crabbiness and quietness. I couldn't say or do anything to bring his spirits up. The work issue was a huge hit, and when followed by losing his dad, the Fisher house wasn't a pleasant place to be. What did I do? I called the people I knew *could* help—I called on the brotherhood. I rallied his driver, his best buddy from Seattle FD, and a dear friend from Austin FD. All three showed up at our door without question and knew just what to do. They had a piece of firefighter artwork for him, and on the back of the framed print was written "Get Your Mind Right." Would I, as his partner, have been able to pull that off without an argument? Doubtful. Was it perfect and needed coming from his confidants? Yes.

I've learned that as much as I want to be the one who can be everything to my partner, sometimes I just can't be. And it doesn't mean I've failed. I need his fire service family to help out when I call. I need to be involved in his work life, I need to know his driver, his coworkers, and their families, so that when the time comes—and it will—I will not be ashamed to tap out and call in the cavalry. I won't be embarrassed to make that phone call and I will know there is no judgment or negative comments.

Chapter 6 — Essential Conversation #3: Handling the Tough Runs

> With these lessons comes the trust of a marriage: I will always do what's best for my partner, even when that means finding someone else who is best for that situation.

Mike does, on occasion, actually want to talk about the tough run. It's then my job to stop everything and really listen, uninterrupted. I'll admit that some of the stuff I hear is pretty horrible and upsetting, in spite of the fact that he screens out the worst details. The very first such dark thing I remember hearing and being shocked by some thirty years ago was of an aid run he'd gone to where the deceased had hemorrhaged his lung into the bathroom sink before dying. I really didn't want to know such things could happen. I remember Mike saying rather calmly that he didn't know what it was at first. I sat speechless and listened and tried to keep my mouth from dropping open. Over the years I've learned to be somewhat thick-skinned and not freak out if I find he's willing to open up and vent. Sometimes I don't think he even realizes he is venting, he's just speaking matter-of-factly. But if he's telling me, he's obviously wanting to get it out.

If I sense pain there, I don't let my firefighter be too vague; I keep gently poking and prodding until I get him to tell me his *true* feelings, which are my primary concern. Never do I give him a forced interrogation into grisly details that he may not be ready to share just yet. The emotions are the key: if a firefighter will express those, then a spouse is better able to comfort.

Countless professionals have told us, and we ourselves have experienced, that men in particular are uncomfortable expressing sorrow, much more so than women. Because they are more reluctant to actually feel and express dark emotions, they assume others are too. Women, however, are much more naturally sympathetic; the area of the brain devoted to emotion and communication is twice the size in a female brain than in a male brain. Remember this, men—you can trust your wife with your feelings, as she has a superhighway of conductivity in this arena.[2] Your sorrow will not make her as uncomfortable as you fear it will. And husbands, your firefighter will very likely need to talk it out, so don't discourage her desires to do so.

If my firefighter wants to talk, I listen. Usually, however, my firefighter doesn't really want to talk about the tough runs *at all*. I then must let go of my curiosity and choose not to feel shut out. He'll let me know he's had a particularly bad night, say briefly why, and that's all I get. So what do I do? He's clearly hurting but doesn't really want to talk about it, at least not to me, so I find other ways to offer support. Mainly, I simply cut him some slack for that day. He *loves* it when I lighten his load by taking something off his plate, like letting him out of that family birthday to stay home and chill instead (never the kids' birthdays of course, just the extended family ones). This is probably his favorite show of support, so much so that he has, on occasion, jokingly attempted to take advantage by *feigning* sorrow just to get out of going (but this I've learned to recognize as well). As always, find the balance.

Another show of support is extra patience. I'll let him be irritable without retaliating, as I know where it's coming from and that it will soon pass. And the more grace he receives from me, the more quickly it *does* pass. Or perhaps I'll prepare his favorite meal, suggest he take a nice, long, hot bath, or put aside the honey-do list so he can simply rest. These are the gestures he treasures when he needs a break. My needs are different, and he in turn reciprocates comfort in my way when I've been traumatized. I like to talk, talk, talk about my pain, and I want him to hold me and just listen. He has mistakenly "just left me alone" in the past, which is the last thing I want. Know each other well, well enough to give comfort in the way your spouse feels truly comforted. That's what love is for, as Amy Grant sings: to "round off the edges, talk us down from the ledges, to give us strength to try once more."

Giving comfort means making someone feel better when they're down and out…when darkness comes…when pain is all around. This could mean a neck rub, a back rub, a foot massage, or really great sex, which is, of course, a favorite. It can mean no pressure to perform today: just relax, I'm here for you, *I'm on your side.*

The impact the car fire (with our children as the possible victims) incident had on my firefighter a few years ago did not surprise me, but his response to the recent near-death of his coworkers did. I thought I'd seen it all after 30 years, but this was not the case. Mike's crew (from a different shift) was involved in a gas explosion that they all miraculously survived. Mike did not witness the explosion and his crew received only minor injuries, but he was extremely agitated and anxious just the same. This surprised me. I wanted to say, "They're fine, chill out," but I didn't. He spent practically the entire day desperately trying to reach them all on the phone, and he nearly threw a fit when no one had the probie's phone number, as it was only his second shift. What did I do on this day to help my firefighter? I stayed out of his way. I let him spend hours on the phone without complaint, then I even went with him when he met with some of them, as he could not be contained. He had to know that they were all ok, he even had to see it. And my quiet show of support was just what he needed from me.

Understanding trauma and its impact on your firefighter is one of the greatest gifts you can give them. Mike and I once attended a speech given by a 9/11 survivor and this point was really driven home for us. She was on the 67th floor of the World Trade Center when the first plane struck. She relayed with graceful, tasteful detail her harrowing eye-witness account of all she went through that day. Our eyes filled with tears as she said, "I can still see their faces—I even made eye contact with them—the faces of the firefighters who were walking past me as I raced down the stairs and they quietly marched up them."

Her struggles with PTSD are ongoing, she told the crowd, and this was made evident to her when her husband brought her to a recent air show. As the planes soared and raced overhead, she was so filled with anxiety that she had to leave, and her husband was irritated by this. At the end of her speech, Mike asked her what her husband could do to help her cope with her trauma. She thoughtfully replied, "I wish he would take the time to do the research, to understand...so that I wouldn't have to tell him all the time what I need and why. This makes me feel selfish, even demanding. It would help me if he would figure some of this out on his own."

Firefighters risk themselves for strangers all the time, they lay themselves down as a bridge from danger to safety. How much more should we spouses be willing to do the same for them once in a while? Understand how trauma impacts the soul and never stop looking for information. Have this essential conversation and find out what helps *your* firefighter handle the tough runs. You are each other's greatest source of strength and encouragement.

A DIFFERENT KIND OF TOUGH

You never know which run is going to be the one that gets to you. The one that makes you realize you are not made of steel and that stays with you at night when you close your eyes. Over the course of a fire service career there will be many of these incidents and, thankfully, most of us develop coping mechanisms to navigate the emotions they cause. Your marriage should be at the forefront of navigating the pain and frustration that develops when tragedy and suffering start to eat away at your peace.

I am grateful to have the wife I do to walk with me through some of the difficult times. There are so many things I wish I had never seen, yet I am proud to have been there to try to make things better. It's the conundrum faced by every person who works to help those who are in trouble. But make no mistake, those bad days have an impact. A loving, caring, understanding partner is the single best balm for the ache that defies description and the erosion that naturally occurs if not attended to. Anne has been that in so many ways. And it has not always been easy, as hurting people tend to hurt those who are closest. I know some of my reactions to what is bothering me have not been kind or reasonable. At times, it's like it is out of my control to soften the tone or consider my words before speaking them. Both firefighter and spouse play a role in navigating the realities of the challenge of tough runs.

DON'T LET THE BAD BECOME SOMETHING WORSE

I will start with the firefighter, as Anne has done a great job on the spouse's perspective. Most important for me to remember is that it is not my family's fault that bad things happen at work. It makes no sense to transfer the darkness that I witness on certain runs to the wonderful light I have at home. That needs to be something you remind yourself of again and again. It is understandable that what you see causes pain, but it simply does no good to react in ways that only increase the problem with your loved ones. That mindset of not allowing stuff we cannot control to destroy what is most precious to us is essential. It breaks this terrible cycle of feeling pain and then inflicting pain and then having to deal with increased pain.

Firefighters also need to understand that not everyone knows exactly what we do. You need to convey some of that, in a respectful and compassionate way, to your spouse. Help them understand why some of the stuff gets to you from time to time. Often, we just assume, because we are feeling it so strongly ourselves, that others around know what we're going through. That is typically not the case, at least not completely. Spouses have their own stuff to deal with in their jobs, family, and personal lives. Make sure you are being clear if you expect your spouse to get it and help.

JUST SAY NO TO DRUGS... AND YES TO SELF-CONTROL

One coping mechanism I have seen has brutal consequences: turning to chemicals to deaden the feelings. Plenty has been written about this so I won't belabor the point. But alcohol, drugs, extreme behaviors, and the like, all end up making things worse. In many cases, irrevocably so. You end up being the tragic example that some other firefighter or police officer will have to deal with as they try to help you on your worst day. If these types of things are how you're choosing to deal with what's inside, please seek help.

Finally, I'd plead with you to get your mouth and your mind under control. It's not OK to be verbally abusive or continually distant from your spouse, even when you are in agony. Some space is just fine. A slip up here and there on how you respond is certainly human. But you control what you say and how you say it. Give your spouse a chance to play the role they were designed to play. It can be the most beneficial medicine you'll ever take.

SAVING THE SAVERS

To the spouses reading this, I offer a simple addition to what Anne has said. You are everything to us. You alone have the access to our hearts to make a difference in calming the terror we sometimes feel. We may put up walls; I pray you seek ways to get around them. We may speak harshly; I hope you'll give us the benefit of the doubt and work to get at what is behind the anger. We may withdraw; I ask you to give us some room to breathe while never giving up on us or letting us drown. You are the savers of those whose jobs it is to save others.

Marriage was designed to be for better or for worse. Some of the tough runs are the worst of the worst. Hand in hand, committed in body, mind, and soul, you can have great success. Take it on together and hold each other up.

NOTES

1. Tom Kenney, "PTSD: A Spouse's Role," *Fire Link*, October 19, 2009, http://firelink.monster.com/training/articles/9167-ptsd-a-spouses-role?print=true.
2. "9 Differences between the Male and Female Brain," Brain Fitness for Life, April 23, 2017, https://www.brainfitnessforlife.com/differences-between-the-male-and-female-brain/.

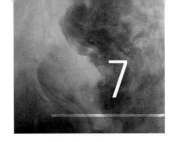

Essential Conversation #4: Dealing with the Fix-It Mentality

I once picked up Mike's bathrobe, which he'd hastily dropped in his early-morning mad dash to the firehouse. Upon doing, so I stumbled across his slippers which lay underneath—his favorite, expensive, soft, comfortable, lamb's wool slippers from Australia. (We have laminate floors, which can get quite cold on bare feet, so Mike wears his warm slippers nearly year-round.) In the half-light of predawn, I noticed something odd about them—one toe seemed to shine. I turned on the overhead light. My eyes widened in surprise and I laughed out loud: slapped piecemeal across the toe of one slipper were layers and layers of silver duct tape! He'd had a small tear that caused the sole to flap, hence the duct tape. If this doesn't truly embody the fix-it mentality, I don't know what does.

Men tend to be fixers by nature, and the best fixers of all? Firefighters. Firefighters (whether male or female) are fixers on steroids. If a building is on fire, they put it out. If someone's hurt or trapped, they rescue them. If a tool breaks, they fix that too. If the rig gets damaged, they repair it. They fix things, they fix people. That's their job. They're hired because they're good at it. If they see a problem, they must solve the problem, any way, anyhow, and often without proper resources or manpower, *very* quickly. They simply get it done (fig. 7–1). They turn chaos into order and restore peace to the scene of tragedy over and over again. They fix people's problems on their worst days because this is what we ask of them. This mentality is so ingrained, so required, that it becomes part of who they are, and they can't help but bring it home with them.

Fig. 7–1. Firefighters need to fix situations that are practically unfixable. Courtesy of John Odegard.

Billy & Teri Goldfeder, Loveland-Symmes FD, Ohio

I guess we should start off with an understanding that this is a second marriage for us both, so a point of order: Billy's first wife was not exactly thrilled with his obsessive involvement in the fire service. So to start off, it is critical that if you are "ate up" with the fire service, , whoever you are dating should do a full and complete size-up prior to saying yes to any further relationship development.

DON'T MARRY SOMEONE IF YOU'RE PLANNING TO CHANGE THEM

Marriage requires a lot of compromise but the best time to talk about what you want in your marriage (home every night? kids?) is before you get married. To be clear, Billy has been an active firefighter since 1973 and has never slowed down:

- He always has and still does have the LSFD fire radios on 24/7/365 regardless of whether it's his department due on a fire or not. He likes to monitor a comfortable 200–300 square mile (mi^2) area just in case, because *you never know!*

- With the advent of the internet, he is hooked up to dozens of other fire department alerts that he is affiliated with across North America, again, *just in case.*

  ◦ Note: You can't always foresee the future, so my health issues and need for sleep led us to the "night tones" compromise:

If he can't respond (you know, if Hawaii is on fire), then that tone gets silenced till morning. The exception is if his son is on duty—we both want to monitor his runs (your kids are always your kids no matter how old you both are).

- He reads every issue of *Fire Engineering*, *Fire Rescue*, and numerous other fire service publications each month, cover to cover.
- He attends every fire training session he can.
- He has a wardrobe that is 98% navy blue fire shirts, some that are decades old, but as he says, "These are classics, they have plenty of life left in them."
- He has a social circle that consists almost entirely of firefighters because nonfirefighters simply don't get it.

We could write volumes about this but hopefully you get the picture. As Ronald (the arsonist) said in the movie *Backdraft*, "The funny thing about firemen is... night and day they are always firemen." Billy takes that to an unusually high level and I love it, and I love him for it. He has the biggest heart of any man I've ever met. But that also means I have to share him a lot, so being independent is a must for a spouse of a firefighter.

I met Billy when I was in basic fire school, so my interest was there and generally, it works, but it doesn't work automatically, and that is the point we want to share with those reading our contribution to this book. Marriage and a successful relationship are not automatic. While the first year or so of marriage often feels automatic due to the new romance, kinda like a new car, ya gotta maintain it to keep it running. We are not kidding. It takes work—a lot of work—and to repeat, *it is not automatic*.

One of the things I "sized up" about Billy is his 100% commitment to the things he loves, whether the fire service or his family. All you have to do is read one of his books or listen to him speak to understand his commitment to and love for his family: his three kids, my two kids (his stepsons whom he treats as his own), his sister, his mom, and of course his beloved "six pack"—our beautiful grandbabies. His commitment to family was extremely important to me, as my commitment is the same.

Billy also knows a lot of people, specifically in the fire service, business, and political worlds, due to his involvement over many years. He can't be close friends with everyone, so he does have a smaller group, an inner circle that he would literally do anything for and they for him. I have seen it and it is a true blessing. I have seen him do whatever is needed for anyone who needs his help—sometimes to a fault, so as his wife, I do my best to give him guidance and advice when I see something good or not so good.

This isn't me sticking my nose into something; this is me looking out for my husband.

COMMUNICATE

We try our best to communicate. We are both very strong-willed people so there are predictable times where we are going to lock horns. I won't BS you: it is not easy and can cause significant stress, but eventually we work it out by communicating. I need time to process and think stuff through when we have a challenge, while he, on the other hand (very much in firefighter/fire chief fashion), wants to fix the problem *now*. Let's evaluate it, let's analyze, let's fix it, and let's move on. Not exactly. I need more time to figure things out, so that can be a challenge but we have learned about each other by communicating and generally how to—and how *not* to—work things out.

Communicating isn't texting and it's not normally electronic. The best is face-to-face. Sometimes, though, putting our issues or concerns in an email or a handwritten letter can be helpful. It allows the other one to read at their own pace and have time to process. It has worked for us.

GO TO BED

Some will suggest you don't go to bed angry when you have a problem. For us that's impossible because while Billy may want to solve a problem, I need some time. So going to bed angry may happen once in a while, but you'll be fine. It can lead to a more thought-out solution. Try not to bring issues up at bedtime (when both of you are tired)—timing is everything.

TIME CHANGES EVERYTHING

One of our favorite bands is a group called Asleep at the Wheel. They have a song called "Time Changes Everything" and that's true about people. The person you marry today will not be the person you are married to 20 years from now (hopefully, they will be an improved version). We aren't talking about physically. Everyone knows that Teri and I are both as insanely attractive now as we were when we got married and will be when we turn 100—that's not the issue. The key is maintaining the love (through communication!) so that you both change as individuals and as a couple together. In a way, you fall more in love. Sound like a fairy tale? Well, it can be, or it can be a nightmare if you don't communicate and simply expect your relationship to be fine without some serious work and effort.

DON'T APOLOGIZE

After all that, from time to time each of you will need to apologize. So why did we write "don't"? Because sometimes words are empty. Make sure they are backed up with a *plan* for action that shows you are sorry or you will make the same mistake again. Actions and heartfelt words will keep the relationship strong.

DON'T COMPETE

The two of you are the two of you. Do what you enjoy and don't try to compete with other couples. It can be challenging sometimes when they have this or do that, but forget that. We need to do what works best for us. We need to buy what we need for us. We need to take care of our family *our* way and not worry about what other couples may have or do. We are us.

SHUT UP

It's not cute when couples bicker in front of others even if you are "just kidding" (because you aren't). You are, as we say in my old New York neighborhood, "bustin' chops" and that's not what you should do in front of others. You want to impact others and their view of you? You don't have to—if your relationship is working, it will show and be very obvious. (You can also just make out. *A lot*.)

FINALLY, A NOTE ON WEDDING PLANNING

Recall how much time you spent planning your wedding. Each and every detail, perfect for this "magical" day. Sure, the wedding day is nice and is the start of a relationship, but that's our point—it's the start. Your wedding isn't your marriage. If people would spend 1% of the energy they spent planning their wedding on their marriage, both short and long term, there would be a lot more happy people in this world and a lot fewer divorced couples.

We understand that firefighters bring their fix-it mentality home with them. But what happens when they do? How do I respond when Mike tries to fix *me*? I'll tell you: with an explosion of rage and offense. Instead of putting the fire out, he's just doused it with gasoline. What works so well in the firefighter's professional life can absolutely backfire in his personal one. Firefighter: no matter how *constructive* your advice may seem to you, if it's coming across as criticism, it will only be *destructive* to the relationship. Criticism is a death blow to a marriage because it kills trust, so be very careful how you speak. Your job is not to fix your spouse, but to constantly edify them. Love builds up; it *never* tears down.

Pretty much all the women I know like to vent, including me. We have powerful emotions that boil and churn within us day in, day out. We feel deeply, we love deeply; we are extremely empathetic and sympathetic. We care—about everything. But these powerful emotions can sometimes turn negative, and we need to let them out. How do we generally do this? We like to talk them out—vent. We feel strongly, and we like to talk about it. This does not as readily occur in the male brain, but in the female brain, as talking releases feel-good chemicals (oxytocin and dopamine) that alleviate anxiety. Because of this, it is therapeutic and healing and natural.

Men don't always understand this. I'll be unburdening my soul when I'll glance at Mike, and catch a blank stare. He'll quickly rearrange his face into a look that implies interest. But I caught him. I now know he's not getting it; however, I do appreciate the effort. Mike knows, as do most men with successful marriages, to let me talk it out, whatever *it* is. He has learned to suppress his fix-it mentality, or the urge to immediately offer suggestions, and just lend an ear, a kind word, and maybe even a hug. How did he learn this? From having the essential conversation on this topic—and it is an essential one, especially for fix-it firefighters. In the past when I'd vent, Mike would try to fix whatever was ailing me, point by point. For example, looking for reassurance about my appearance, I'd say something like, "I feel fat. Do I look fat to you?" Mr. Fix-It would reply, "If you feel fat, get on the treadmill." Small explosion. I'd say, "So-and-So hurt my feelings when they said such-and-such." Mr. Fix-It would reply, "You're too sensitive; you need to not overreact; and why do you care what they think anyway?" Small explosion. Instead of getting the sympathy I was after, I've just been told my reactions are incorrect.

Through the conversations that resulted from these exchanges, Mike now understands that I am not a project—I don't want him to fix me. I am merely looking for comfort and support. When does the offense occur? When he crosses the line from fixing the problem to fixing *me*, as no one likes to be made to feel inadequate, especially by the person who is supposed to think you're wonderful. Mike realizes that my feeling deeply causes me to wound easily, so he must tread lightly. And some things cannot be fixed; it just helps to talk them out then let them go.

Pretty much all the men we know speak as little as possible, especially about their feelings (there are exceptions and, as with everything we discuss, apply this information to your situation and personalities). If they're saying it, it must be important—because the unimportant isn't worth the trouble of vocalizing. For women, it is usually the opposite; we may be saying something we don't really mean—like *does this make me look fat?*—when what we really mean is *do you still find me attractive?* Or we may just want to purge. This can be confusing to a man; does she want my advice or doesn't she? Do I try to fix this or just listen? And how am I supposed to know?

There are, in fact, occasions when I do want his advice. These are entirely different than venting sessions, and the experienced, seasoned spouse learns to differentiate. I'll say, quite clearly, "What should I do?" That's the green light for his explicit opinion. Mike is very wise, and I do value his guidance when it's geared toward fixing problems, *not me*.

On the flip side of this hard conversation is Mr. Fix-It's point of view. Mr. Fix-It can, on occasion, be right and fair in his attempts to fix me. I've learned that sometimes I am overreacting; I can be too sensitive and I've had to face this. And I do need to get on a treadmill if I want to feel better about my appearance. Mr. Fix-It doesn't intend to offend me; he's genuinely trying to help, however insensitive his suggestions may at first appear. He'll point out that I often try to fix him and he doesn't get offended, and he's right. Because of our combined experiences and perspectives, I've become stronger and he's become more tactful. By communicating his help in loving ways and listening without fixing, Mike has allowed me to grow without crushing my tender heart. And my patience with him has helped him to become an incredibly empathetic partner at home and a leader at work.

IT'S NOT ABOUT THE NAIL (EXCEPT WHEN IT IS)

The problem of the fix-it mentality within marriage once came up at, of all places, one of Mike's teaching engagements for local firefighters on a topic totally unrelated to marriage. The class went off without a hitch and the students listened politely and were engaged and interactive, which is what a teacher always hopes for. After the class, several of the firefighters rushed up to Mike excited to speak to him. He assumed they wanted to discuss his class, and the majority did, but to his surprise, some had another agenda. They couldn't *wait* to show him a video clip and were actually giddy with excitement to do so.

Later that evening Mike came home to me. I asked him how his class had gone. He brushed off my question with a "Fine, good," then abruptly changed the subject by saying he had to show me a video clip. He actually giggled as he said so. What in the world did this video have to do with anything and why was he so determined to show it to me? The clip started playing on his laptop and he continued to chuckle as I began to watch.

The scene opens with the profile of a pretty young woman talking to her boyfriend or husband, who is sitting next to her on a couch, listening patiently. She's describing pain, headaches, how she feels all this pressure, and how she's confused by it all. She then turns to face him, and we see that she has a *nail* in her forehead. The boyfriend or husband eyes the nail and says, "Well, you do have a nail in your forehead," pointing out the obvious.

She tells him, "It's not about the nail," and "Quit trying to fix everything," and "You just need to listen."

She goes on with her diatribe, which includes her frustration that all her sweaters are snagged, and he is noticeably frustrated and struggling with what to say. At some point he cautiously says, "That must be really hard."

She gratefully responds, "Thank you!" as this was all she wanted from him.

By the time the clip is over Mike is practically rolling on the floor with laughter. This is what the firefighters had wanted him to see, as they knew about my marriage columns. I was only slightly amused. In fact, I felt more irritation than anything, and I wanted to defend my sex from male insensitivity. The fix-it issue is a major source of conflict in marriage and I was ruffled by this mockery. Do men really think women are so stupid as to not recognize when we have a *nail* in our forehead? Do they really believe all our problems are that easily identified and resolved? Enough so to actually make a professional video on the topic?

I begin to remind Mike of the differences between men and women; women are sensitive creatures with incredible abilities to identify and sympathize with others. The emotion center of a woman's brain (the hippocampus) can be up to two times the size of a man's; therefore, we handle and process emotion at a much deeper level. We deeply care about people and are impacted by other's emotions. Scientists call this "mirroring emotions," or the ability to *feel* what our loved ones are suffering.[1] This incredible sensitivity allows mothers to pick-up on the non-verbal cues of infants. Men do not have this gift.

I remind him too, that a woman's communication center of the brain is typically bigger than a man's, and we like to talk—about everything. It literally makes us feel better. A sensitive man understands this and learns to just listen, to let us vent, to hear our pain without inserting their advice on how to fix the problem. This is not something to be made fun of or belittled.

I'm just getting started. I go on to remind him that we have raging hormones that fluctuate as much as 25% in a month, while men's hormones only fluctuate about 2% a month, so of course we get a little moody and emotional; we can't help it sometimes.[2] Why is that so funny? Our huge levels of estrogen render us capable of loving so deeply in the first place; they should thank their lucky stars that we do! This supply of estrogen makes women desire relationships and connection. Just because our neurological reality can be like the weather and hard to predict, and theirs can be like a mountain that slowly erodes over time, is no reason for them to belittle us.

Chapter 7
Essential Conversation #4: Dealing with the Fix-It Mentality

I then reminded him that testosterone is the hormone that influences energy, aggression, competitiveness, and courage, and high levels of it can *compel* men to fix and compete. And men usually don't have the same brain capacity to just ride out emotion; in fact, they are often very uncomfortable with sadness. The typical male brain reaction to an emotion is to avoid it at all costs. Because of this, men often view people as projects. Tears represent chaos and must be stopped.[3] I believe this is why women tend to be better care givers, because we *don't* try to fix people.

Finally, I admit it: I'm insulted by the implication that the solutions to all women's problems are that easily identified, that we just need someone to point out these obvious answers. They aren't and we don't: life is complicated and truly hard to understand at times. Problems don't typically have easy solutions, and it's absurd to think so for even one second.

A few days after viewing the nail video, I was venting to Mike about an ongoing issue I'd had. He'd heard it before and, like the good husband he is, has listened patiently and empathetically on this topic many times. I wait for his learned response of support, but instead, this time, he simply stated that it was my own fault, as I knew what I needed to do but simply refused to do it. At first I was taken aback by this gentle but firm truth. I started to feel slightly offended, but then I suddenly realized that *he was right*! He was absolutely right that time. My problem *was* a simple one with a simple solution that I'd chosen to blow completely out of proportion!

I began to rethink the nail video, and a smile came to my face: sometimes it *is* about the nail! Men and women do often think differently and see things differently, but this is a good thing! Our strengths and weaknesses complement each other, and if we can just have a bit of humor combined with a dose of tough love, we can learn to live peacefully together and fix things at the same time.

The *very next day* after my epiphany, Mike and I were speaking together at a chaplain's conference on firefighter marriage. The crowd was responsive and wonderful. At the end of our session, a group gathered around us with comments and kind words and questions. A woman approached me and said she'd like my advice about something. I told her I hoped I could help and asked her what the issue was. With frustration in her voice, she asked, "How do I get my husband to quit trying to *fix* everything?" This was the moment I decided to officially add the conversation about the fix-it mentality to both my book and my PowerPoint, complete with the nail video. This is a conversation that firefighter couples must have, as it is obviously an issue. Not only did this wife ask me how to stop the fixer, but I've had many husbands ask me how to better "fix." The topic (and the video) invokes both laughter *and* meaningful discussion wherever we go.

Firefighter husband, remember the slippers with the duct tape. Just as lamb's wool slippers are the softest, so is a woman's heart. A caring, gentle woman is a tremendous source of warmth and comfort to a firefighter who must face a dangerous, cold, cruel world. Please do not slap duct tape on your wife when she comes to you with her broken pieces. It is an ugly, harsh, abrupt fix. It can cause scars. It can result in hardness. Partnerships require tenderness, understanding, and support. Instead, gently sew from within, from the heart, with words that are reassuring of your unfailing love. This will take longer, but it will help keep them beautiful both inside and out. And spouses, give your firefighter a little slack when they break out the duct tape or point out the nail. They are, after all, just trying to help. It is their very nature to do so. And sometimes, a quick fix is *exactly* what you need.

The nail video accurately portrays the lunacy of a woman wanting to *just talk* about all the pain from the nail in her forehead, but men have their own idiotic ways too. Men usually do not try to talk about their issues—instead they tend to ignore the nail altogether or, even worse, find ways to work with or around it. That's where spousal support comes in and can be incredibly beneficial: we can point out each other's nails and help each other not only recognize them but remove them as well. We all want unconditional love and acceptance from a spouse, not criticism, but sometimes tough love is needed to get to the truth; that we do indeed have a *fixable* problem, or in other words, a nail in our forehead. This is not criticism, it is help.

HOW TO IDENTIFY A NAIL

1) You know it's a nail when action is required

So many problems in life cannot be fixed. These issues may impact us, but we cannot impact them, so in these instances, talking or venting really is all we can do. For example, we have no influence over the economy. We have very little impact on government, except for our vote. We can't choose family or coworkers. We can't make people treat us well. We can't control the weather or unexpected illnesses. And we often don't have much say in job requirements or salary. These issues are not nails.

But what can we fix? What *is* a nail? If you're feeling depressed about your weight, you *can* get on a treadmill and eat less; this is a nail because you can take action. If you're feeling overwhelmed by too many work hours—cut back, say no—action required, problem fixed. If there is distance forming between you and your spouse, make time for each other, go out on more dates, or stay home together—action required. If you're avoiding a project and it's stressing you out because it's not getting done, watch less TV, spend less time on the computer, and get going—action required. If *your* kids (not other peoples' kids) are out of control, practice some tough love and set up a proper routine with boundaries and expectations—action required. These are the types of stressors that impact us that we can indeed impact in return. The problems they cause can be solved with *action*.

2) You know it's a nail when two or more people witness it

Let's face it, sometimes we ignore our spouse. We don't always hear what they say. Husbands, admit it, you often put on your "wife ears." And wives, you'll shut down if you feel your husband is being insensitive, whether he actually is or not. Or sometimes we've heard it so many times that "it" just becomes background noise, something we tune out. As a result, we don't always take our spouse's word for it, as we may have become a little biased against their opinion on certain issues. We do this because, again, we want unconditional love and acceptance, warts and all. Our flaws are part of the package that our spouse simply must endure whether they like it or not.

But we know we have a nail, or a fixable problem, when someone else *other* than our spouse sees it too. This is Old Testament stuff: Two or three eye witnesses are required to convict someone of a capital crime—two mind you, as one may be biased.[4] For example, if both your spouse *and* your mother think you're awfully tired, heavy, or unhealthy looking, maybe you should see a doctor. Or if both your spouse *and* your coworker observe that you're always at the fire station, maybe you're working too much. Or if both your spouse *and* your child detect undue anger, perhaps, firefighter, you've been traumatized on a run and should seek some wise counsel. Multiple witnesses confirm the obvious, and we can no longer deny that the nail exists.

3) You know it's a nail when you keep applying the wrong medicine

This is the oft-quoted folk definition of insanity: doing the same thing over and over again, while expecting different results. Like Norman Bates famously says in *Psycho*, "We all go a little mad sometimes." We do indeed exhibit madness when we repeat the same old bad habits but expect things to change. We try to fix the nails in our lives with the wrong medicine, which only increases the pain, turning the nail into a railroad spike.

For example, we despair that our clothes are getting too tight. We feel sad and depressed, so what do we do about it? We eat to feel better. Wrong medicine! We can't sleep, we're always tired, we're falling behind at work and are totally stressed out. So what do we do to eliminate the stress? We binge watch TV all night to escape reality. Wrong medicine! Our finances are a mess, we're in debt over our heads, which is pulling us ever downward into despair. So how do we lift our spirits? We go shopping (online or in the store) and buy something new. Wrong medicine!

This is insanity, and sometimes it takes a good long look in the mirror to see how we're causing our own problems. Bad habits compound our issues, causing even more pain. It's like taking pain pills for the headaches the nail causes instead of removing it! Masking the problem typically leads to more severe problems or damage. To change bad habits, they must be replaced with good ones, and a spouse can help with this if we let them.

Marriage is meant to be a support system, not just any support—but the best available. If you'll listen to each other, you'll better recognize your nails and be able to remove them together. Firefighter, be careful to not fix when listening is what is really required. Learn what your spouse will hear as criticism and never *ever* say those things. Firefighter spouse, remember that sometimes they just can't help themselves. Fixing is their language of love; it isn't meant to be criticism. Have this conversation and be aware of each other's tendencies and deepest needs. It will save you from many unintentional, unnecessary arguments.

FIXER-UPPER?

Boy, do we hear this a lot: "How do I get my firefighter to just listen and stop trying to fix everything?" The good citizens of our cities, towns, districts and areas call upon us during the worst days of their lives with one pleading desire: Fix this! ("this" being whatever problem initiated a call to 9-1-1; fig. 7–2). We react and bring every bit of experience, training, and knowledge we possess to do just that. It seems to be stamped somewhere in our genetic code to fix the problem, as though it is a mandate, a calling. And it is very appreciated when played out on the streets assigned to our firehouse and crew.

Chapter 7   Essential Conversation #4: Dealing with the Fix-It Mentality

Fig. 7–2. Mike's crew shown fixing it. Courtesy of John Odegard.

    Similar appreciation is not often found, however, when we roll on home to our spouse and apply the same fix-it mentality to them. Typically, our spouses do not want to be fixed. They are not looking for instant answers or a step-by-step guide on how to solve the things they are bringing up. In most cases, our spouse is simply venting and just wants an ear to listen. Even in those instances where a solution is being sought, they usually are wanting to collaborate on a solution and just talk about it first. Our rush to handle the problem we've just been told of in a "9-1-1 manner" creates its own kind of emergency and one that need not happen. Your spouse is not an addict with a needle in her arm needing rapid intervention; she is your soulmate needing your tender, patient time and attention.

BEING WHAT YOU HATE

For my fellow fixers, can I remind you of something? We hate micromanagers. We hate them. Bosses who are in our business telling us how to do things instead of taking care of their own worlds are about as frustrating as it gets. Whether you like to hear it or not, the step-by-step fix-it mentality is a form of micromanagement that is just as bothersome when on the receiving end. The presupposition here is that you have all the answers and that your spouse couldn't possibly contribute to what you know. It presumes an arrogant posture that they haven't considered the profound wisdom you are currently extolling from on high and that the answer is really self-evident. Step back a bit and see how you feel when it's you getting this treatment. I bet you don't like it.

I have found that Anne's initial queries on things are part of her process of working things out and usually do not require much input from me. She wants an ear, to vent frustration at times and to hear her ideas in her own voice and gauge their effectiveness. Many times, the simple act of just listening helps her sort through her own ideas and come up with a great solution. In the situations where venting is what she needs, my interrupting with a plan only exaggerates the frustration she feels, even if the plan is a good one. You can save a ton of bad days by just doing what your mate needs and not what you want to do.

For the spouses, I would again encourage you to recognize that we are only trying to help. It is not a personal attack or indication that we think you don't know how to solve problems. Our world is fixing problems and we try to get to it as quickly as possible and in the most effective way. It is a natural reaction that is never intended to be disrespectful or dismissive. Be sure and remind us of what it is you're truly looking for in these types of conversations to give us the best chance of doing the right thing.

BEING A TEAM IS THE TRUEST FIX

The closing thought I'd leave here is to always seek what the other person needs most. This takes time and is a part of growing together as a couple. Lots of trial and error and, hopefully, a heavy helping of grace will be needed to mature together. But that's what you want. You joined hands and pledged yourselves to each other to be a great team. You recognized you were better in unison than you were going about on your own. As with so many areas we will discuss, this is one of those key times when collaboration is the answer. It will fix the fix-it problem and make for much more harmonious decision-making as you proceed.

NOTES

1. Brizendine, *The Female Brain*, 122–123.
2. Ibid., 4.
3. Ibid., 123.
4. Deut. 17:6.

ESSENTIAL CONVERSATION #5: KEEP YOUR FIRST FAMILY FIRST

It's 1990 and we are a young couple with two small boys ages three and two. We're a tight little family, close, and newly returned to Seattle (my hometown) from serving in the US Air Force at England Air Force Base in Alexandria (LA). Mike has spent the last year working as a corrections officer in Tacoma as he tests for his dream job—Seattle Fire. Then the day finally came, the day that would change both his life and mine for the next three decades: the day he received his acceptance letter from the Seattle Fire Department.

Firefighting has been an amazing career with incredible highs, incredible lows, and a few surprises as well. And sadly, we have watched many of our peers along the way lose their first family to the phenomenon that was then unforeseen by us—the second family.

Firefighters form very tight bonds. They are a team, a team that puts their lives into each other's hands. They tend to be like-minded with similar interests, even similar senses of humor (such as gallows humor.) They share inside jokes and experiences that often cannot be spoken of with those who weren't there. This bond is a brother/sisterhood that can become as strong as blood. Like soldiers in the trenches, as firefighters battle the forces of death they discover this:

> *It awakened in us a strong, practical sense of espirit de corps, which in the field developed into the finest thing that arose out of the war—comradeship.*
>
> —Erich Maria Remarque, All Quiet on the Western Front

The strong affection firefighters feel for one another is a *good* thing. It helps them do what they do—and love doing it. But through our years in SFD we've watched this bond become a problem if and when it begins to take *precedence* over a firefighter's first family—their spouse and children. And this can be a hazard to a firefighter's marriage, one that we've had to address ourselves over the years.

Rita Brunacini, wife of legendary, much-beloved, and painfully mourned Chief Alan Brunacini (Phoenix FD), stated this problem best when she complained to Al, "I think you love the guys at Station 1 more than you love me."

Al, whose sharp wit and sense of humor is well known in the fire service, quickly replied, "Yes, I do, but I love *you* more than the guys at Station 2."

We chuckle at this funny response, but the underlying jealousy is real enough, as I myself have experienced on many occasions. Sometimes I've felt left out, as I cannot compete with sirens and flashing lights and "dragon fighting." The urgency of life and death often trumps the needs of a sheltered wife. Or so it can seem. My firefighter, who adores me above all else, has sometimes felt his loyalties being torn as he is desperately needed at both home *and* at the firehouse. The balance between the two is something that every firefighter couple must find, and to do so I suggest having this essential conversation: *How do we keep our first family first?*

How does a spouse *feel* that they are number one over a profession that is elite, exclusive, and vital to the existence of our communities? Being needed is a compelling force. Needed to save lives, needed to help other firefighters, and needed because you're excellent at something brings a sense of pride. Firefighter couples can find themselves grappling with this force when it begins to call a firefighter away from his home *too much*, infringing on his first love, his first family. Boundaries must be set between the two families. If they are not, and a firefighter begins to neglect his wife and kids, then be prepared for the consequences—losing your first family to the second one. And let's be real, firefighters: as compelling as your firefighter family may be, you can't live with them 24/7, and you will, eventually, retire.

RED FLAGS OF WARNING

To keep "us" first, my firefighter husband and I begin with this step, which is to recognize the "red flags of warning" as I call them. These are indicators that Mike and I are beginning to drift as a couple into marital discord, warnings that I am beginning to feel second.

Red Flag #1: We're not talking much about deeper issues, which is one of my vital needs to feel connected. Conversation becomes shallow, minimal, and perfunctory.

Red Flag #2: Mike begins to develop a hair-trigger temper and is easily irritated by me or the kids. I begin to feel as if I'm walking on eggshells.

Red Flag #3: When he's here, he's not really here. So distracted is he by the projects and demands piling up that he has trouble concentrating on just me. When one or more of these red flags begin to creep into our marriage, which is more precious to us both than anything else, we drop everything and take immediate action before things get worse, before I begin to truly resent his profession.

Every couple is as unique as the individuals that comprise it, but here are a few tips that my firefighter and I have found after decades of SFD experience to help keep the lines between first and second families firmly in place.

Make sure time together is focused

The call of the firehouse is a strong one—it can be felt, even at home. My firefighter is sometimes so busy as to be perpetually distracted and this I can and do resent at times. So what I need from him—what I've told him that I need—is his *undivided attention*, not just his physical presence. Time spent in meaningful conversation is the best gift you can give your marriage. Conversation that includes feelings and deepest thoughts creates an atmosphere of affection that bonds a couple as nothing else can. The firefighter schedule can disrupt this bond with its erratic in-and-out pattern, but time usage can repair the dangerous trend to drift apart, which seems to be the number-one reason behind firefighter divorce. Begin that reconnection by having affectionate conversation, as women, in particular, need it every single day.   This means no cell phones at the table. Look at me when I'm talking and actually listen to what I'm saying—and this I'll know you're doing when you give the appropriate responses. He never means to neglect me—I am undoubtedly first in his heart and he tells me this all the time. But he *shows* me I'm first when he puts all distractions aside and turns his full attention to me—if just for a little while—each and every day. This makes me *feel* first.

Run every major decision by your spouse

Doers get asked to do—a lot. And let's face it, most firefighters are doers. My husband gets asked to do major projects (and some small ones too) almost every day. When he gets asked to take on the big ones, he must do this—tell me about it *before* he makes the decision to accept.

It's easy to be compelled by the urgency of the firehouse to jump in headfirst. But here's the dilemma: these projects will ultimately impact me, so it's only fair that I be in the loop. He has learned to find out as much as possible about all that is involved and then share it with me. I am then shown that my opinion matters. I have buy-in when I have a say in the decision and am thus better prepared to accept the intrusion into family time if and when it comes. And this makes me *feel* first.

Give up the extras

This applies to both the firefighter and the firefighter spouse. Firefighter, if you want to take on that compelling extra work project, then cut back on time with the crew off duty or give up a hobby or two. Mike quit some real sports teams and some fantasy ones along the way to make room, as time with me was not going to be the cost of doing business. It was hard for him to give up the fun stuff, but time with me is fun too. It's hard for me to be the "bad guy" by asking for these sacrifices, but we both believe it's never a bad thing to fight for your marriage. Pursue and plan time together as intently as you would a hobby—make your spouse your hobby like you did when you were dating—and your best, most enjoyable times will be spent together, not apart.

And if you don't want your firefighter working two or three jobs because you're drifting apart as a couple, then firefighter spouse, you may have to give up those extra dollars. I've had to ask myself many times over the years if I'm compelling my husband to work every overtime because I'm griping about money. If it's all about the toys, marriage and family may start to feel second. Working every available overtime, second jobs, third jobs—ask yourselves, firefighter couples, is it really worth it? Those toys aren't much fun when divided in divorce.

Our experience in working with many couples has convinced us that one of the deadliest enemies of marriage is the pursuit of wealth. It has been my experience that many of the firefighter couples who struggle the most are the ones who spend too much time *working*. How can this be? Making money is a good thing, right? Absolutely, money is necessary to support a family, to meet needs. But the scales tip from necessary to dangerous when it is pursued *at all costs*. If the desire to build a fortune becomes stronger than the desire to build a family, if every minute is spent working two, three, four jobs on top of firefighting (which is already exhausting) then all energy will be used for financial gain. This leaves no energy to spare for each other. There are only so many hours in the day. Some *must* be saved for preserving the most precious possession you will ever have: a loving, intimate relationship.

At one of our speaking engagements, I met a firefighter who told me that he used to work *five* jobs to support his family. They had plenty of money, but he and his wife had drifted apart and decided to separate. In a last-ditch effort, he cut back to three jobs, and to accommodate, they lowered their standard of living. The marriage not only recovered but was flourishing. "Less is truly more," he told me. "No amount of money would have been worth it if I lost my wife along the way." This is but one of the stories I've heard, stories of marital successes when the workload was whittled down and stories of marital failures when it wasn't.

If you can't find the time to connect as a couple in conversation and lovemaking, in date nights and shared interests, cutting back on work may be the answer. Ask yourselves, do we want more money or more time? Fortune or marriage? More work or more *life*? Is the ladder of success leading to divorce? If so, consider another ladder. And redefine success.

Live without the extras—extra play, extra money, extra achievement. If your marriage is top priority, you don't *need* the extras. Mike choosing time with me over expensive toys or hobbies or building a fortune makes me *feel* first.

Say no more often

This has been an interesting phenomenon we've discovered through our long experience as a firefighter couple: the more he says no, the more I say yes. How does this work? Because of the aforementioned tips, my firefighter now routinely runs every major activity or project by me first. And he has learned from this to say no if it comes at too high a price, which is even more time away from me (or the kids when they were home.) He has said no to promotions, to extra income, even to many social outings with the crew. In fact, he became so good at the *no*s that my trust in his discretion and obvious deference to me flourished. In the early days, his knee-jerk reaction was always yes, and my knee-jerk reaction was always no. But with healthy conversation, sacrifice, and balance, the opposite has come to pass. He gets lots of yes answers from me now, wholehearted ones, confident ones, because his *no*s have made me *feel* first.

Time invested in the kids keeps the marriage strong.

Many women, including firefighters, have shared with us that the more a man is involved in childcare, the better his wife feels about the marriage. And I would heartily concur. Mike never sacrificed our children on the altar of his career, and this is a hard thing for an ambitious man to do. He deliberately chose not to become an officer until our boys were nearly graduated from high school because he knew this would take him away too much. He was there to coach the boys' teams, to volunteer at their schools, and to give them guidance all throughout their teen years. His presence kept them on the straight and narrow; they never got into any trouble. They went on to college with both academic and leadership scholarships. It was only once they were firmly on their way that he chose to promote.

He sacrificed his own hobbies to help the boys find their talents, and this devotion to them communicated his love for *me*. I never had to feel like a single parent, which is a common complaint in the fire world. He met one of my deepest needs by being an *excellent* father. His time with our children made me *feel* first.

Do fun stuff together

Because firefighters are active, athletic thrill seekers by nature, the firehouse is a major draw, because that's where the "fun" people are. On duty, off duty—firefighters just want to have fun. They see some pretty grisly stuff and play is a healthy way to combat that stress. If home is no fun, then the firehouse may be a more desirable, even irresistible place to be. The scales then tip in favor of the second family.

To maintain my status as number one, I've determined to be just as good a "playmate" as the firefighters. Mike and I do fun stuff all the time, and the more fun I try to be, the more he'd rather be with me than with anyone else. We laughed and played and found fun activities we both enjoyed when we were dating—who says it has to end after the wedding? Dr. Willard F. Harley writes, "Couples don't need to share all interests, just a few if they hope to communicate well.... Marriages with completely divergent interests may survive, but I've never seen one that's flourishing."[1]

Play is very important to high-stress workers, especially to men, as it is typically listed as one of their top marital needs. Make an effort to have fun as a couple, and firefighter, make sure you let your spouse be your best friend. Don't spend all your most enjoyable moments with someone other than your spouse. The more Mike has sought me out as a companion, the less I have resented his time spent with the crew. In fact, I encourage it (fig. 8–1)—their close bonds improve morale—and morale brings courage, confidence, and enthusiasm to the firehouse. This keeps my firefighter not only happier, but safer. But his willingness to ultimately spend more time playing with me than with them has made me *feel* first.

The firefighter has the unique blessing of having two families; most professions *do not*. The second family will be there like no other in a firefighter's darkest hours; not only his, but his entire family's. Fire spouse, know that if your home should ever catch fire or your child is ever hurt, your firefighter family will practically drive people off the road to get to you. Their devotion to one another is absolutely, hands down, one of the major perks of the job. It is a blessing; let it be a blessing and not a curse by finding that elusive balance between the two. You *can* have both, but only if you strive to keep your first family first.

Chapter 8   Essential Conversation #5: Keep Your First Family First

Have that conversation as a couple and find out what works best for you; what makes *your* spouse *feel* first?

Fig. 8–1. When first and second families blend beautifully. Mike's crew and their spouses.

Brian & Janine Root, Seattle FD, Washington

First and foremost, if anyone out there thinks their career is the highest priority, then they're destined for marital stress (or worse) rather than marital bliss. Your partner should be at the top of your list. The fire service is incredibly rewarding, but to make a marriage work well in this profession, you need to be able to do a couple of things right. You must communicate and you must share this job with your loved one. Share the good along with the bad.

It should come as no surprise to anyone that maintaining a happy, solid marriage involves good communication. Within the fire service, there are certain things that inherently aren't discussed in great detail on the home front. There is this notion that we must be the fixers, the stoic ones, not permitted to show emotion. As a relatively new firefighter, I learned this the hard way with my girlfriend, Janine, who is now my better half (and still my girlfriend!). We all have had those calls that shake us to our core, and I'm willing to bet that we each remember in detail the very first one that put a scar on our heart.

Firefighters are wired to put themselves in harm's way to shield others. I believed this included my suffering in silence from a difficult call that I had been on so that Janine would not have to be exposed to the details that haunted me. I thought I was doing the right thing by keeping it to myself, but I was, instead, alienating the people around me who cared the most. During the days that followed, I went through many different emotions. At times, I was angry. Other times, I was depressed. I lost sleep. My appetite was minimal. I didn't find joy in the things that I used to. I was just plain moody as I went back and forth through all those emotions.

It wasn't until Janine reached her limit and put me in check that I realized how I was acting, and more importantly, how it was making her feel. I thought that by telling her I had a rough call I had given due diligence to communication—wrong! You see, I was trying to be a robot and not show the extent of the emotional toll. That painted an unclear picture of what I was trying to process. Therefore, from what Janine could see, there was no good explanation for my behavior. Had I just taken the time to admit that I wasn't coping very well, there would have been some added leeway given at home. It doesn't mean that I needed to share every little detail of the tragedy with her; I just needed to let her know I was having difficulty shaking off a call.

The other part to making a fire service marriage work is sharing the good stuff about this job. Who better to share it with than the one person who has promised to be by your side? I have previous experience in a marriage that did not survive the fire service. I take blame for not ensuring that the wonder of this organization was being shared at home. I selfishly kept the experience to myself. The result was that my ex became resentful toward the fire service and everything associated with it.

I was lucky to find love again and vowed to be a better communicator and share more of the great things about the fire department. You'd be surprised how meaningful it is, on occasion, for your spouse to be invited to dinner at the firehouse. As a couple, we try to do things with other couples in our "fire family," everything from crew Christmas parties and camping trips to concerts and sporting events. It allows your spouse to form bonds with other spouses who can relate to the stressors that come with a fire service marriage. The chances are that your spouse doesn't have a work environment where people are that invested in one another. When you include your spouse and share your fire family with them, it greatly decreases the chances they'll feel they are competing for your time and attention. When they realize there is no competition for you, that they clearly hold the number-one spot on your priority list, the relationship becomes rock solid and *everyone* wins!

TOO MUCH OF A GREAT THING

Being a member of the fire service really is a privilege. Most of us feel incredibly grateful to find our calling in something so honorable and sacred and rewarding. The job truly has a pull that brings out the very best in us and has the potential to elevate in ways that few things can. That's the good side of the fire service.

The other side comes around when we allow the fire service to consume us and become something it was never intended to be. The compelling nature of this work and the people who are drawn to it can obscure the other things in your life that should be priorities. It's in this dangerous area that something so amazing as the calling to be a firefighter can become something sad and destructive.

Our feeling on where family sits in relation to firefighting is simple: family comes first. Period. End of discussion, no debate. When you commit your life to your spouse, that act puts them ahead of whatever profession you choose. When you decide to have children, that addition creates new souls who are now a higher priority than your job as well. Your vows to your spouse and the children that may result trump your firefighting career and even your firefighter family. That is our belief and all that follows stems from that idea.

YOU SHOULD ENJOY BOTH, BUT SOMETHING WINS

There is certainly room in this discussion to acknowledge that we have time for both home and work. That's a reality of life as you must make a living and not everything has to revolve around your home. But this discussion centers on what comes first, what gets your best. Your spouse and family need to know that they are your top priority and if push comes to shove, it's the fire service that gets shoved, not them. All too often it's the other way around as the family gets sacrificed on the altar of serving the citizens.

I BELIEVE I CAN FLY...

Early in our married life, Anne and I were stationed at England Air Force Base in Alexandria (LA). Because of some work I'd done, I was rewarded with the opportunity to fly in an F16 fighter jet. Needless to say, this was a dream come true for many of us and most will never get that chance. When notified, I was elated and couldn't wait to tell Anne. That she was less thrilled than I is an understatement. Immediately we were at odds over whether I should do this. I couldn't believe it! Her fears were me being up in a jet and something going wrong. We had two small boys and the possibility of them losing me was very real and utterly terrifying to her. I tried everything I could to persuade her that this flight was safer than my driving to work, but she kept asking me not to do it. When I mentioned this to the guys at work, they were beside themselves. As you can guess, I was the subject of significant abuse when I told them my decision on the matter: if Anne is not comfortable with it, I will respectfully pass.

Whew, was my manhood called into serious doubt. The ending of this story is a good one, as Anne is a reasonable person. She had time to process it and go through all the safety precautions in place for a flight like this. When all was said and done, she decided that if I really wanted to do it and felt it was as safe as possible, I had her blessing. I took the flight and it was epic.

The moral here is that our decisions are made together. All of them. When work intrudes on our home, work loses. There is obviously some balancing that must happen as we all must make a living. But here is the truth, as hard as you may find it to believe: if the fire department started having a destructive impact on my family, I would quit. I would earn a living in some other way. What good is a successful career in the fire service if it costs you your family? What good is the accolade that comes with being a "hero" when those closest to you suffer?

YOU DA MAN?

One of the most compelling things that can happen is to be viewed as the go-to guy/gal on your job, always available and always willing. It's tempting and very satisfying to be praised in that way and our culture celebrates it. We'd encourage you to consider being the go-to for your spouse instead. Be the one who is always willing and available to meet the needs of your family. What would home life look like if that were the case? How close would you be to those who truly matter most?

Seeking to balance work and home life is a challenge that must first be acknowledged and then dealt with. A family life that is prioritized will help you be better at work and give you some leeway at home when you do make a mistake. It is very doable, but it must begin with you deciding what matters most. Hopefully that answer is your first family.

NOTE

1. Willard Harley, *His Needs, Her Needs: Building an Affair-Proof Marriage* (Ada, MI: Revell, 2002), 65.

Part III

How to Live with a Firefighter

THE FIREFIGHTER AND TIME OFF

Even if no run occurs in the night, firefighters don't really sleep on shift. They are in a constant state of waiting preparedness or tangible anticipation, listening for the bell to go off in the night. When it does, the jarring shock of bright lights and loud noises disrupts their natural circadian rhythm which, as described in an earlier chapter, is the body's internal clock that regulates sleep and wakefulness in a 24-hour period. It has been suggested that continual circadian rhythm disruption can lead to an absolute mental and physical breakdown, just like perpetual lack of sleep in a combat soldier. The two-day break has been established as the minimum amount of time needed to break the circadian rhythm disruption and return it to normal, with four days being optimal. The break also returns the firefighters' cognitive abilities to 100% (after only two nights, circadian rhythm disruption can reduce cognitive abilities to 65%). Without the break, firefighters would have difficulty performing their job tasks for more than one or two years. Additionally, ongoing research is finding that continual circadian rhythm disruption has also been linked to serious illnesses, including cancer.[1]

What this all means is that your firefighter physically *needs* the multiple-day breaks. You can build up sleep deprivation, but just as our bodies can catch up on hydration and nutrition, so too can they catch up on sleep. Getting caught up on sleep is one of the best ways to heal and recover from psychological stress. For these reasons, it is paramount that the firefighter couple does not load days off with extra chores. Instead, let this break include plenty of time to catch up on sleep and exercise. This is vital to the firefighter's well-being and improves job performance. And it helps your marriage too.

INDECISIVENESS

Life-and-death decisions are mentally exhausting, more so than any other type of decision-making. At the end of a 24-hour shift, the firefighter can be so mentally drained that they have nothing left. How does backlash affect marital decision-making? Spouse: "Honey, what do you want for dinner?" Firefighter: "I don't care." Not caring what we have for dinner is a very common issue in our home, one for which I no longer care either. Firefighter spouse, you will find that you are forced to make many small decisions without the input of your firefighter, it's just a reality. But here's the upside; if he has no opinion when asked, then he must accept whatever is handed to him without complaint. This is a light-hearted example, but a real one. Know that your firefighter *does* care, they're just sometimes too drained to say so. Do not mistake a temporary detachment as a sign of relational apathy—it's mental fatigue, nothing personal. More importantly, make all big decisions when your firefighter is well-rested so that they truly will be made as a couple; this will save a lot of grief down the line.

TRAUMATIC EXPERIENCES

There are reasons firefighters are given lots of time off, the primary being to reset circadian rhythm disruption after a 24-hour shift, as was mentioned before. Again, it takes at least two nights of unbroken sleep to reset just one night of disruption. Without a reset, a firefighter can develop serious sleep deprivation, which predisposes them to becoming a stress casualty (one who develops PTSD or ASD). But sleep is not the only issue.

When police officers are in a gunfight, the International Association of Chiefs of Police (IACP) protocols state that they are to be given three or four days off to recover. If they return to duty too soon and another incident were to occur, they would be vulnerable to severe psychological damage. Many are given *more* than four days if needed. Time off is vital for their mental and emotional health. During a fight-or-flight response, as much as two-thirds of brain function is shifted from the prefrontal cortex (PFC) to the limbic region for the purposes of survival. The prefrontal cortex controls emotion; those suffering from post-traumatic stress disorder (PTSD) show reduced activity in their PFC, resulting in chronic heightened anxiety. Without time off to literally shift gears, both police and firefighters are more prone to suffer from PTSD.[2]

Firefighters must play the long game, sustaining mental acuity and physical strength for a career's worth of service. Time off is meant to be spent resting, recovering, and repairing from sleep disruption as well as the influx of stress hormones and their resulting effects.

For the firefighter spouse, time off means two things; don't resent it, and don't overfill it. Most people do not get as much time off as a firefighter, spouses included. There have been times in my life that my job as a mom went 24/7. Then I added more work on top of that to bring in extra income. When I overloaded myself, I found that I began to resent my husband's seemingly easier schedule. Something had to give and we both determined that it wasn't going to be our marriage. I let go of some of my extra commitments, which meant a little less income, but it was worth it. Sometimes less really is more: more time spent together and greater appreciation for his schedule.

I also came to understand *why* he needed those days off by facing the realities of firefighting. For the first few years of his career I was an ostrich; I ignored the dangers because they frightened me. But for the firefighter marriage, ignorance is not bliss. As was discussed in chapter 2, parasympathetic nervous system (PNS) backlash is a direct result of exposure to danger and must not be misconstrued as laziness. Therefore, do not expect your firefighter to fill that extra time off with a longer honey-do list. The firefighter's honey-do list should be the same length as anyone else's.

If your firefighter works in a busy house, try to avoid that second job or having him work every available overtime. Again, time off is for physical and mental recovery from the most stressful job on the planet. If your marriage is struggling, perhaps trimming the household budget is the answer. An overworked, overtired firefighter can be tough to live with. A happy marriage and solid family are far more precious than any amount of extra money.

Sam & Amy Hittle, Wichita FD, Kansas

We have spent 17 of our 21-year marriage in the fire service. The fire service profession adds its own set of unique challenges for any marriage. We do not, by any means, have it all figured out, but we have learned and grown along the way. We have come to understand that one of the most important things we can do is make our home a place to decompress and rest after a long shift. One of the ways we accomplish this is to make communication a priority.

Samuel has often come home frustrated with the job, fatigued, disheartened by a traumatic call, or energized from a great fire. In any of these circumstances, he can come home and share the events of his shift with me and know that he is safe to do so. We have noticed many of his brothers want to spare their wives the gory details of the job. Most of the time it is probably with good intentions as a way of protecting them. In some circumstances, the firefighter is still processing their own feelings and some may even feel that their wives wouldn't understand these feelings. Most women, however, and especially in the case of firefighter wives, are much stronger than people give them credit for. Their capacity for understanding and empathy is unmatched if allowed the opportunity. The wisdom, reassurance, and support a firefighter can get from his wife can have enormous benefits to their health and well-being.

Communication is a two-way street. Having a trusted person to open up to and being allowed to process feelings out loud in a safe space, free from judgment, has been one of the greatest gifts of our marriage. It's not enough, however, to have the willingness to communicate. You must also have the opportunity to communicate. This is going to require making a focused effort to set aside time with each other. It may just be a cup of coffee on the deck, a walk around the block, or even an official date night but we try to start the week by looking at our calendars and blocking out time for each other. We both have full-time jobs, Samuel is very involved in his department as well as the national fire service, and we have two kids with busy schedules, but we make it a priority to set aside several opportunities during the week for time together.

SLEEP

A team of psychiatrists followed a veteran unit of soldiers into battle during the Korean War. The unit was attacked and fought an intense battle in which they were successful but a counterattack was sure to soon follow. As they waited for the enemy to retaliate, the psychiatrists were astonished to see that the soldiers were all falling asleep! They had to be awakened by the company officers, repeatedly.[3] PNS backlash is so powerful that the body literally collapses, even in the trenches of war. It has to shut down in order to replenish its energy supplies, and the only way to do so is to sleep.

Studies have shown that sleep deprivation is equivalent to being legally drunk.[4] When lives are at stake, a sober, rested mind is paramount to making critical decisions. Airline pilots and doctors are required by federal law to get a certain amount of sleep as life-and-death decisions are part of their jobs.[5] The same should be true for firefighters, as their profession was recently deemed to be more stressful than any other. Sleep is the warrior's best medicine. It resets hormone levels and clears the mind like nothing else can. Unfortunately today, we're seeing high levels of PTSD in returning Iraq and Afghanistan troops, and the reason may partly be due to technology. Instead of returning to the barracks to sleep, American soldiers are playing video games, watching movies, and surfing the internet till all hours of the night. The human brain typically can only hold seven new items of information at a time (hence the 7-digit phone number.) More than seven items causes the brain to process the information into long-term memory, and it must stay active to do so. Media stimulation is a means of escape as addictive as drugs or alcohol. But if the mind is active, the body is not truly at rest, and this is alarming as sleep deprivation is a contributing factor to developing PTSD.[6]

In the firefighter home, sleep needs to be as serious an issue as it is for the soldier, as firefighters fight a war that never ends (fig. 9–2). Make sure plenty of time is allotted for rest after 24-hour shifts and limit late-night media binges—though they may appear restful, they are not. Only during sleep is the PNS truly ascendant, enabling the body's energy reserves to be restored for the next battle and giving the mind a chance to process the day's influx of information.

Fig. 9–2. Mike's crew, Ladder 5. Time off needs to be well spent to show up ready to rock 'n' roll with these guys.

EXERCISE

Stress hormones are meant to be used in fight-or-flight. If called to a fire but no fire occurs, the hormones will remain in the system. Adrenaline at home impedes sleep and impacts relationships as it can lead to inappropriate harshness or aggressiveness. Running, lifting, and playing sports burns off excess adrenaline, allowing a firefighter to relax and shift gears. Workout routines are often the first thing to go in a busy household, but they should be the last. It's worth every penny to set aside a small portion of the family budget for either a gym membership or home equipment. Exercise is just as vital as restorative sleep in maintaining a healthy firefighter and a happy marriage. The topic of exercise warrants its own chapter, so it will be explored in more depth in chapter 10.

FOOD AND WATER

Stress hormones can wreak havoc on the digestive system. Adrenaline and noradrenaline suppress the digestive system, while cortisol increases the appetite.

As a result, firefighters have high rates of indigestion, acid reflux, and spastic colon. Studies show they take a disproportionate amount of antacids.[7] Food has a satiating, calming effect, which is a nice break from anxiety. This can be a good thing—in moderation. Unfortunately, placating stress with food can be just as addictive as alcohol or media consumption, which may then lead to another firefighter killer; obesity. Couples need to be aware of this tendency and have strategies in place to combat food cravings when planning family meals. Try to stock the shelves with healthier choices and limit bedtime snacks, as a full stomach also interferes with much-needed sleep.

Water is a firefighter's friend both on and off the fire scene. In an incredible article, "Hydration and Firefighter Performance," written by our good friends Mike McEvoy and David Rhodes, they assert, "Firefighters working in gear can lose two percent of body mass within 30 minutes." This impact could result in a "loss of ability to concentrate, lowered alertness, feelings of tiredness, and headaches." They go on to say that firefighters are often significantly dehydrated. The solution? Water: compared to all other beverages, water remains the "quintessential fluid" for maintaining hydration. It is essential for keeping the body hydrated before, during, and after a fire, and increases both performance and recovery. Studies show that dehydration also increases levels of cortisol, the stress hormone, which suppresses digestion.[8] Firefighter spouses, encourage your firefighters to drink lots of water at home, as this will help to keep them healthy, less prone to injury, and fit for duty.

SEX

PNS backlash can have many negative repercussions on both the firefighter and their marriage. But there is, however, one major perk: it may increase the sex drive, especially after a life-threatening situation has been successfully resolved.[9] Can you think of any better way to fill the two- to four-day break? I can't! Leave lots of time for "afternoon delight"—it's good for you both and keeps your marriage strong and intact.

A GOOD THING GONE BAD?

As Anne has mentioned, we really like our work schedule and consider it a blessing. As with most positive things in life, it also brings aspects that can be truly problematic. Great food is wonderful. Getting fat is not. Having good friends is one of the best things in life. Getting so involved with others that you grow distant from your spouse is not. Making some extra cash working overtime is a great way to fund some fun stuff for your family. Working too much defeats the purpose altogether.

So it is with our schedule. Not all firefighters work 24-hour shifts but it is very common in paid and larger departments. Most of what we speak about can be adapted over to other types of schedules, including volunteers who roll out at all times of the day and night as well as working their regular jobs. The point of this discussion is to recognize that the challenges of our job mandate some down time. Not getting any breaks to refresh, recover, and realign is a recipe for serious disaster. I've watched this play out in many of my comrades' lives and there are two primary observations I feel are critical for couples. One is for the firefighter and one for the spouse.

For the firefighter, there is this whole mentality of invincibility. We feel we can go and go and go and catch up somewhere down the road. But that rummy feeling we have at the end of a busy shift is our body needing sleep and restoration. Even though we get used to fighting it off and moving on to job number two, picking up the kids, or doing those other department projects, the impacts are real. As Anne has so aptly described, the effects of seeing some truly terrible stuff, making so many critical decisions, and the added sleep deficit diminish our ability to do normal stuff in the normal way. That's what the time off was intended to address. If utilized properly, the rest time will allow you to function at your best when functioning at your best is absolutely required.

As gung ho, Type-A folks, we really believe we can make it all work. And we're interested in so many things that the free time allows us to pursue lots of different stuff outside the fire department. My plea to my fellow firefighters is to commit to making rest a priority. For most of us this must be a deliberate, planned effort as it does not come naturally. When we abuse our system with too much activity, the tendency is to then pursue artificial means of compensating. Alcohol ends up being the remedy of choice in most of the instances I've seen and usually adds to the problem instead of solving it. Some firefighters I know have felt the need to use stimulants to get them through their second job and then stagger back to work in hopes of a slow shift. That's a risky bet and one with terrible consequences if lost. Don't take the risk. Make rest a priority and be thankful we have a work schedule that allows us that break.

THE GREEN-EYED MONSTER COMETH

For the spouses, I find it interesting how our schedule can become a source of jealousy. Because our work is concentrated in big blocks of time, the time off can appear bigger than it actually is. I would respectfully encourage all spouses to give their firefighter the benefit of the doubt and plan some rest on those days off. Whether you recognize it or not, the temptation is to see that time as something that can be filled with whatever chores or projects that need doing. Some of that is fine, but must include dedicated time to recover from the very real impacts of doing our work, both physically and mentally.

HEY DUDE, WHERE'S MY CAR?

A friend and mentor of mine, Captain Bill Gustin of the Miami-Dade Fire Department, relayed a story that is helpful in illustrating how much exhaustion can affect us. Bill is truly one of the most committed firefighters I have ever met and someone who has dedicated his life to the fire service. When we were chatting about this issue of being gassed when our shifts were over, he heartily agreed. He described getting off work and heading to a local bagel shop to get a bite to eat before heading home. After spending much too long at the counter trying to make a decision over a simple thing like getting breakfast, he headed out to the parking lot. In that short span, he could not remember where he parked his car. This type of thing, he continued, was typical after work as the brain is taxed. He did not want to make decisions, couldn't remember things that were normally easy, and was overly blunt in his conversations. Don't be surprised if this is the way your firefighter comes home. It is not personal.

The desire on the part of the firefighter to fill up off time with interesting, productive stuff is a natural thing. Our spouses seeking to get things done and using our block of time off to make that happen is also normal and, at times, necessary. What is critical is that each of you recognize the toll our work takes on the mind and body. Getting rested and returned to top form has got to be part of the plan and viewed in the appropriate way, or really bad stuff is on the horizon. Plan it together and you can make this good thing truly wonderful.

NOTES

1. Grossman and Christensen, *On Combat*.
2. Mark C. Shantz, *Effect of Work Related Stress on Firefighter/Paramedic*, Eastern Michigan University School of Fire Staff and Command, Fourth Class, research paper, August 2002, 18.
3. Grossman and Christensen, *On Combat*, 16.
4. "Study: Sleep Deprivation Is Like Being Drunk," (video) *USA Today*, https://www.usatoday.com/videos/news/health/2017/11/07/study-sleep-deprivation-like-being-drunk/107442456/.
5. Sarina Houston, "FAA's Final Rule for Pilot Duty and Rest Requirements," The Balance, November 17, 2017, https://www.thebalance.com/faa-final-rule-pilot-duty-and-rest-requirements-282927.
6. Grossman and Christensen, *On Combat*, 26.

7. P. H. Barnes, "The Experience of Traumatic Stress Among Urban Firefighters," *Australian Journal of Emergency Management* 14(4), summer 1999–2000, 60–64, http://eprints.qut.edu.au/2121/1/2121.pdf.
8. Paul Huljich, "Drinking Stress Away: The Link between Water and Stress Reduction," Stresspandemic.com, October 17, 2012, http://www.stresspandemic.com/blog/drinking-stress-away-the-link-between-water-and-stress-reduction.
9. Grossman and Christensen, *On Combat*, 276.

Exercise

Because of the high-stress world of firefighting, we firefighter couples have certain problems that plague us more than the general population. The dangerous aspects can lead to physical problems such as injury, sleep deprivation, and disease. The traumatic aspects can lead to mental and emotional issues including depression, anxiety, attention deficit, and addiction. And all of these impact marriage. For the firefighter couple, exercise is *absolutely paramount* because it counters these problems in the healthiest way possible. Exercise should be a top priority in the fire household, but unfortunately in a busy schedule, time to exercise is usually one of the first things to go.

Firefighters are typically given a few days off to recover from a 24-hour shift. What do these go-getter types usually do with that extra time? What do budgetary demands often *compel* them to do? Fill it with extra work. Resist this temptation firefighter couples, because it can add to marital strain, not alleviate it. To keep your marriage and your firefighter healthy for the long run, you must allow adequate time for exercise.

In our household, my husband has carte blanche when it comes to an exercise budget and allotted time. We have an entire home gym that we both use on a regular basis. If Mike wants a new piece of equipment, we find the money, we make the room. If you don't have room in your home for a gym, then go for daily walks or runs (which are free) or purchase a gym membership (which is worth every penny).

How much exercise is good for the firefighter and the firefighter marriage? Moderate aerobic exercise for 30 minutes, three days a week is good and very doable. Moderate intensity falls in the range of 65%–75% of your maximum heart rate (which is based on the rough formula of 220 minus your age.[1]

But how much exercise is *best*? According to John J. Ratey and Eric Hagerman, MD in their book, "Spark," the optimum benefit of exercise both physically and mentally occurs with an aerobic workout of 45 to 60 minutes, six days a week. Four of those days should be on the longer side, at moderate intensity, and two days should be on the shorter side, at high-intensity, or 75% to 90% of your maximum heart rate. For a 45-year-old, the theoretical maximum heart rate would be about 175 (220-45). If you calculate 75% and 90% of your maximum, the lower and upper limits for a high-intensity workout are 131 and 158. On the shorter, high-intensity days, adding resistance training maximizes overall fitness. The high-intensity days should not be back to back as both your body and brain need recovery time. We recognize that opinions are all over the place on what's the best exercise routine. The above is a great place to start, but find what works for you and what you'll actually do.

BENEFITS OF EXERCISE

Many of the issues that plague firefighters are primarily due to the stress hormone cortisol. "Chronically elevated cortisol levels are both inflammatory and catabolic and cause a myriad of disorders including: thyroid and metabolic dysfunction, cognitive decline, low serotonin levels resulting in depression, altered glucose metabolism, elevated lipid levels, increased blood pressure, low melatonin levels resulting in altered sleep patterns, musculoskeletal issues resulting in difficulty recovering from exercise and possible subsequent injuries. Cortisol levels are also related to mental acuity and can factor into degenerative diseases such as dementia and Alzheimer's."[2] That's the bad news. But here's the good news: if firefighter couples will commit to exercising at least moderately (but hopefully both, as moderate is good, but intense is best), all of the negative effects of cortisol can be countered. Overall mental and physical health can improve, which ultimately will benefit your marriage as well.

Exercise makes firefighters fit for duty

Not everyone has to be fit to do their job, but firefighters do. The fitter they are, the better they are. He must be strong. She must be able to lift people. He must move fast. And she must be able to do this job for a very long time. Strength training increases muscle strength. Aerobic exercise increases agility. More strength equals greater performance, and more agility means greater likelihood of returning home, injury free.

Movement and stretching removes lactic acid build-up in muscles from exertion, reducing the likelihood of muscle soreness or tears. To be fit for duty, firefighters *must* exercise. To live long and grow old together, spouses, encourage your firefighter to do this; even better, exercise with them, as it is more fun to run or walk with a partner. Make it a part of your daily routine, or better yet, use it as bonding time to strengthen your relationship. Physical fitness also makes you both more attractive—not just fit for duty, but fit for passion.

Exercise counters stress

Stress is an emotional and physical carnivore. No one encounters more stress than a first responder. They have most of the pressures of an office job plus dangerous and traumatic situations. Since stress is such a monster and so unavoidable, we as firefighter couples must address this issue and learn how best to cope with it. As stated before, exercise is a fitting answer, as stress is a physical response and exercise provides a physical cure.

Why is stress so harmful? The stress response (fight or flight) is intended to be a good thing. Stress calls into action the most powerful hormones in the body and scores of neurochemicals in the brain for the purposes of survival. When faced with a stressor, say, a fire or a car wreck or even a personnel conflict, the amygdala sets off a chain reaction. Within 10 milliseconds, the adrenal gland releases norepinephrine and epinephrine (adrenaline), which increase the heart rate and the respiration rate and causes some muscles to contract to prepare the body for action.

A third hormone is also released—cortisol. Cortisol thickens the blood enabling it to clot faster if an injury should occur. This will help prevent the body from bleeding out. Cortisol does more. It signals the liver to make glucose available while at the same time blocking insulin receptors at non-essential tissues and organs. It does this so that the body will receive less glucose and the brain will receive more for the purposes of quick, reflexive thinking. Cortisol simultaneously burns and restocks energy stores by converting proteins into glycogen and begins the process of storing fat, which is most often stored around the belly. If the stress response is triggered *but not used* (i.e. a false alarm or a personnel conflict) belly fat piles up and muscle tissue breaks down.

Cortisol affects brain function as well by shifting glucose away from the thinking parts of the brain to the limbic or action center of the brain. If continuous or chronic, this can starve cognitive areas of the brain and lead to damage, such as short-term memory loss, over- excitability, or the ability to deal with stress in the future. For the firefighter, whose stress is ongoing, all of this seems to be an inescapable hazard of doing business.

Cortisol makes the brain rigid and less flexible to the point of cannibalizing itself. It inhibits serotonin (the peaceful hormone) and slows growth factors, which can cause the brain to get stuck in destructive patterns of thinking. If overtaxed by chronic stress, the brain becomes hyperaroused. Ongoing hyperarousal can eventually cause a break with reality as the mind becomes overridden with fear. Chronic stress is a problem for many in our society, including firefighter spouses. We spouses, after all, have to worry about the safety and well-being of our beloved firefighters.

Enter the miracle of exercise. Exercise counters these negative side effects of the stress response. Activity fires up the recovery process in our muscles and in our neurons, not only healing the body and mind but making it *stronger* for future stressful situations. With exercise, the body becomes calmer, fitter, and stronger, rendering it less easily triggered into the stress response. See it as an actual inoculation or immunization against stress. This is true for all of us, firefighters and firefighter spouses alike. It increases blood flow and glucose use and production, which are the two essential ingredients for cell life, enabling the body to keep growing healthy new cells. The plasticity of the brain allows it to mold and change and even rebuild damaged areas, and exercise speeds up this process. And energy production becomes more efficient as exercise increases the need for fuel in the body *without* triggering the negatives of cortisol. With less cortisol in your system, your body will burn fat instead of storing it.

Michael & Tracy Dulas, Seattle FD, Washington

Back in about 2002, I was at one of my heaviest weights ever. I was traveling for work Monday through Friday, eating out for every meal, and stressed from my new job and lots of deadlines. We were newly married and my wife Tracy had just finished school and was also starting work. We decided to get road bikes and train for the Seattle to Portland bike ride to lose weight and have some fun. Some neighbors stopped by and asked if we rode and said they did triathlons. Turned out they did Ironman Triathlons. We thought they were nuts.

Fast forward a year: we decide to start running together. Neither of us ran much, so we signed up for our first 5K together. That progressed to signing up for a 10K, then a half marathon, then a marathon. We both used the same training plan. We would train individually while I was on the road and then it was a great thing to come home on the weekends and do our long runs together. Running, more so than bicycling or swimming, allows you to just talk to another person. It was a great way to make sure we kept our pace aerobic (a goal on your long run) so we could catch up on everything. When the going got tough, we would encourage each other. As we progressed through our plan, every week brought a distance that we had never run before and a huge sense of accomplishment! While we were apart, having the same plan helped to keep us accountable. We would encourage each other to get the training runs in so we didn't fall behind.

We then went on to start swimming and signing up for some triathlons. This progressed to us eventually racing the Ironman Triathlon together. The Ironman is a 2.4 mi swim, 112 mi bike, and 26.2 mi run. Training gave us a common goal to work toward. It gave us quality bonding time as well—what else were you going to do but talk to the other person? It was a shared hobby, so we could relate to what the other was doing. We understood when the other person was suffering, or having a bad day, or just plain old tired. We delighted in each other's successes. We laughed together when we went to a nice restaurant in San Diego after a race and devoured the appetizer platter in about 30 seconds and the waitress just kind of looked at us in shock. Most of this training process was new to us. Things like running 16 mi, running in the dark on trails after work, riding our bikes to new places we had never been, swimming 2 mi in a lake, watching the sun rise. Sharing in a new, novel experience brings you closer as a couple. Doing this also allowed us to witness each other's accomplishments.

It was actually all of this physical activity that led to me becoming a firefighter. I was bored in my other job and wanted something more active. I wanted a meaningful career, to be able to help people.

Exercise is a great way to deal with the strains of being a firefighter. Even after being up all night on the aid car, I force myself to go to the gym for 20 minutes. It can be anything, I just listen to my body as there is no goal on these days. Sometimes that means I am on the elliptical at level 1. Some days I decide I want to push it. It is always easier to get some sleep on days you have exercised. On the weekends when everyone is home in the morning, I try to go to the gym for a quick workout first, which allows me some time to decompress and be a better father and husband when I walk in the door.

Reflecting on all of this, I can see what a positive effect this shared experience has had on our marriage. We have family and friends who always comment on how goal oriented we are. They notice that we simply say we are going to do something and then we do it. I think much of this derives from setting those goals together for our races early in our marriage. Another benefit is that it brought us close. We got married very young—we met in college when she was 18 and I was 19. We married after we both graduated college. The time we spent together allowed us to talk through any issues we had. Usually it was done while we had endorphins coursing through our system so I think that put us both in a better frame of mind. The goal-oriented nature of our training has also allowed us to see how you get from A to Z. You don't think you can run 26 mi, but it becomes more manageable when you see you just have to run 6, then 7, then 8, and so on. We learned how to plan. That helped us take the same approach when I wanted to become a firefighter, and when my wife got her National Board Certification.

We are now entering our 18th year of marriage. Throughout all of the craziness of having two careers and being very involved parents of two kids, we found that our relationship had maybe taken a bit of a back seat to us being a family. It wasn't in trouble, just a bit out of balance. We turned back to where it all started: going on bike rides. It has been great time spent together and we have relearned old lessons on how to improve our marriage.

Exercise counters depression and anxiety

Depression is serious business, as it can lead to health issues and even suicide. The World Health Organization (WHO) lists depression as the leading cause of disability worldwide ahead of coronary heart disease, any cancer, and AIDS.[3] According to the National Institute for Mental Health (NIMH), 6.7% of all American adults (18+ years) had at least one major depressive episode in 2015.[4] Women are two to four times more likely to experience major depression than men, and are also more likely to experience panic attacks, panic disorder, and depressive disorder, according to a study by the WHO.[5] In 2015, more than 44,000 people committed suicide.[6] A 2012 study by the Centers for Disease Prevention and Control found that while men had higher suicide rates overall, women working in "protective service occupations (e.g., law enforcement officers and firefighters)" had the highest rates for their gender (14.1 per 100,000).[7] About 17% of

Chapter 10 — Exercise

American adults experience depression at some point in their lives. And every 17 minutes someone in our country commits suicide. Tragically, we are seeing more and more firefighter suicides.

Chronic stress is the leading cause of depression, and firefighting is one of the most stressful professions in the nation. Chronic stress leads to elevated amounts of cortisol, which, if unused, can damage the brain and effect how a person thinks and feels. Brain-derived neurotrophic factor (BDNF) protects neurons against cortisol damage in areas of the brain that control mood, such as the hippocampus. Cortisol over time decreases levels of BDNF, and autopsies on the brains of suicide victims revealed significantly lower levels of BDNF. Cortisol can also cause an imbalance of the three "feel-good" neurotransmitters: dopamine, serotonin, and norepinephrine. A deficit of any one of these can lead to depression.

Hormones have a powerful influence on mood as well. Women are two to four times more likely to suffer from depression and anxiety than men because of massive hormonal fluctuations. After puberty, a man's hormones fluctuate approximately 2% a month, for life. On average a woman's hormones fluctuate 25% a month, and they can do so even more during pregnancy, postpartum, and menopause.

Too much stress also causes anxiety, the evil twin of depression. Firefighters can be both depressed and anxious at the same time. Anxiety is fear—fear is the memory of danger. If unchecked, the fight-or-flight response can lead to being *trapped* in the memories of fear, tricking the body into believing the danger has not passed, and this may lead to ASD or PTSD. Cortisol damages the prefrontal cortex (PFC), which controls the amygdala. If the amygdala is not controlled, hyperarousal results and can snowball. Sufferers of PTSD show a smaller than average PFC.

Here is the beauty of exercise: it elevates endorphins, raises BDNF, and regulates *all* of the neurotransmitters targeted by antidepressants, not just one. Exercise helps to balance fluctuating hormone levels, making them not as drastic or severe. As little as 10 minutes of vigorous exercise can immediately improve mood and vigor. Every 50 minutes of weekly exercise correlates to a 50% drop in the odds of being depressed. And only exercise alleviates sensitivity to the physical arousal of anxiety by burning off stress hormones. Exercise literally snaps your brain out of the downward spiral of depression or anxiety by *naturally* restoring the proper balance of chemicals needed to do so.

Exercise spurs PFC growth, (which houses memory) helping the brain to remember "the good stuff" instead of being trapped in the negative, fear-inducing thoughts spurred by chronic stress. Out with the negative thoughts, in with the positive ones.

Exercise strengthens mental acuity

The stress hormone cortisol temporarily increases norepinephrine, which arouses attention, and dopamine, which sharpens focus. This is necessary during the fight-or-flight response to be able to perform scary tasks such as fighting fire or public speaking (which is ranked as the #1 fear.) The bigger the threat, the more the fluctuations, and the bigger the fluctuations, the more likely an imbalance will occur. If cortisol is left unchecked in the brain, cognitive skills cease to be sharpened and eventually begin to erode resulting in attention deficit. With attention deficit one can become unfocused, unproductive and unmotivated. And decision making becomes tough.

Exercise increases dopamine and norepinephrine levels the same way that stress does, *minus* the cortisol. These two neurotransmitters are the leaders in regulating the attention system. Exercise accomplishes this by spurring the growth of dopamine receptors in the basal ganglia which is the area of the brain responsible for attention (this is what the drug Ritalin does.) Exercise increases norepinephrine which calms an overactive cerebellum to decrease fidgetiness and noise distraction.

As stated before, exercise increases the volume of the pre-frontal cortex by helping it grow back to its original size. The brain is always trying to rejuvenate itself by producing new cells, a process known as neurogenesis. Experts say that stress exerts a profound effect on neurogenesis, decreasing the rate of brain cell proliferation. It also can prevent new cells from being "plugged in" within the required 28 days, so the new brain cells die. Exercise not only spurs neurogenesis, but helps the brain to immediately "plug in" the new brain cells. In other words, move it or lose it.

The PFC regulates mental acuity. A diminished PFC can lead to inability to concentrate, difficulty remembering new information, and procrastination. But with exercise, sharpness and productivity return, and with them, the ability to make good decisions in a timely fashion.

Exercise helps fight addiction

High stress can cause chemical imbalances that lead to depression and anxiety. Because of this, the firefighter is more vulnerable to addiction than the general population. In an effort to cope, they can often choose the wrong medicine.

Alcohol, drugs, food (especially carbs), videogames, gambling, and shopping all have the same biological effect—they stimulate the *nucleus accumbens*, or the reward center of the brain. These addictions (as well as sex and nicotine) accomplish this stimulation by boosting dopamine levels in the reward center, which makes us feel good. Raised dopamine levels lead to "cravings" for more and more—and this may have some relationship to the development of addictions.

Sex, for example, raises dopamine levels 50%–100%, which is a healthy increase. Cocaine on the other hand raises dopamine 300%–800%—an unhealthy increase. If raised to unnatural extremes, dopamine receptors become damaged over time and the activity must be increased to get the same feeling as before. Abuse of dopamine can also actually shrink gray matter, which is responsible for self-control. The addiction then becomes "reflexive," overriding the thought process.

For the firefighter, stress itself can become an addiction. The 'stress junkie' gets a burst of cortisol which also quickly raises dopamine levels to induce pleasure. But stress, as any other drug, quickly causes an imbalance. It becomes harder and harder for the stress junkie to find enjoyment in day-to-day life as their threshold for excitement grows higher and higher. As a result, home life may begin to seem rather dull and unfulfilling. The stress junkie firefighter might then become addicted to the rush of being at the firehouse and begin to neglect their family.

For any addict, sudden withdrawal is extremely tough, even dangerous. It tricks the body into survival mode, in which it literally cries out for more dopamine, and this can be overwhelming, even crippling. If dopamine is not made present by some other means, addicts will almost certainly relapse.

Since addiction is so powerful, how can we ever fight it? *We can fight it with exercise.* Exercise is not only a legitimate antidote but also the inoculation against future relapse. Exercise raises dopamine levels the same way that addictions do, but to natural, healthy levels. It also regrows gray matter, the part of the brain that regulates self-control. Addictions form superhighways in the brain—the quickest routes to pleasure. Habits are formed. But exercise helps build synaptic detours *around* these well-worn connections, enabling the addict to find alternate routes to reward. Thus, better habits can be formed. Just 10 minutes of exercise have been shown to curb an alcoholic's craving. Just 5 minutes helps some smokers fend of nicotine cravings for up to 50 minutes. Dopamine produced during exercise blunts cravings and produces more dopamine receptors, which allow the body to absorb it more quickly. Exercise actually restores the body's natural levels, which brings balance to the reward center of the brain thrown off by addictions. It strengthens self-control *and* curbs cravings at the same time.

Exercise helps the firefighter sleep

Nothing is more healing to the highly stressed than a good night's sleep. But for the firefighter, often full of adrenaline and cortisol, sleep can be elusive. Exercise is effective in burning off the stress hormones that keep firefighters awake (but do not exercise heavily right before bed). Stretching (or yoga) before bed is also highly effective as it releases lactic acid and ammonia build-up in muscles that causes tightness.

Both exercise and stretching raise dopamine, serotonin and norepinephrine levels, the three "feel-good" neurotransmitters that counter anxiety. With a more peaceful, easy feeling—the firefighter is more likely to sleep naturally, thus again avoiding addictive sleeping pills and alcohol.

Exercise helps fight disease

Stress is damaging to our health. The stress hormone cortisol clamps down on the immune system, which, if chronic, can render us more vulnerable to disease.

Stress and inactivity have also been linked to arthritis, chronic fatigue syndrome, fibromyalgia, and other autoimmune disorders.

Let's look again at the beauty of exercise, a simple way to prevent some pretty serious illnesses. Exercise optimizes energy usage by triggering the production of more insulin receptors, which means better use of blood glucose and stronger cells. Lower blood glucose means less diabetes and healthier cells means less disease.

Doctors are now recommending exercise to their cancer patients as part of their treatment (as it also counters depression.)

Studies have shown that women who do not exercise have an increased risk for developing breast cancer. Both men and women who exercise regularly have a 50% less chance of getting colon cancer. And active men over age 65 have a 70% lower chance of developing the advanced, fatal form of prostate cancer. Take note—firefighters are prone to these types of cancer.

And exercise is doubly good for the heart as it increases cardiovascular health and burns off cortisol, which increases belly fat.

Exercise strengthens the firefighter marriage

This is perhaps the best reason of all to exercise: it is yet another tool for the firefighter couple to use that may improve, strengthen or even help save the most troubled of marriages. Exercise counters depression and anxiety, which are relationship killers. (Note to readers: Mike and I are strong advocates for trying natural cures first, such as exercise and improving communication. However, it is important to consider all possible answers, or a combination of answers, when tackling these insidious foes. Antidepressant and antianxiety medications are valid, and when prescribed, have been truly effective and critically important to many people struggling with depression and anxiety.)

Exercise strengthens mental acuity—the enemy of motivation and productivity. Decision making becomes easier and procrastination diminishes, helping the firefighter couple to tackle problems head on and keep the household running more smoothly. Exercise can help fight addictions, which may become all-consuming, even to the exclusion of marriage and family. And exercise helps firefighters better fight the diseases that threaten to keep them from growing old—with the ones they love.

THE PERFECT "FIT" IS TOGETHER

This topic seems self-evident for firefighters and should be a natural part of being prepared for the rigors of the work. Our hope is that you'll read these words and feel inspired to take fitness on as a challenge for both of you. As with everything we write about, doing things as a team enhances your connection, optimizes valuable time, and allows you to have the same priorities.

In the outstanding book *Spark: The Revolutionary New Science of Exercise and the Brain*, John Ratey and Eric Hagerman relay the findings of research indicating that exercise enables the body to better utilize energy. The process of exercising triggers the production of new insulin receptors, and they stick around to continue the enhanced efficiency.[8] Ratey says purposeful exercise has wide-ranging impacts, including

- Increased social confidence
- Reduced anxiety
- Easier to fight off depression
- Improved focus
- Easier to avoid unhealthy addictions
- Better decision-making ability
- Healthier babies
- Longer lives[9]

HURTS SO GOOD

One of the biggest fears spouses have for their firefighter is that they will be hurt at work. Ensuring every possible measure is taken to see that they exercise and maintain a high level of fitness will greatly enhance their chances of coming home in one piece. Level of fitness has a direct impact on firefighter safety. Here's a few reasons beyond the typical stuff related to cardiac health:

- Fit firefighters think more clearly.
- Fit firefighters do not fatigue as quickly.
- Fit firefighters can more readily compensate when deficiencies occur.
- Fit firefighters sleep better.
- Fit firefighters are less prone to injuries and their secondary impacts.

Every one of these benefits has a direct impact on being safe at dangerous events. They directly impact how the individual will perform when seconds count and mistakes are often deadly.

LIFTING WEIGHTS OF MANY KINDS

Finally, it is critically important to recognize that working out will help in warding off depression and anxiety. By the time you get done with our book, you're going to be schooled in the facts surrounding the challenges of this calling. Its impacts can weigh heavily on the hearts and minds of your firefighters. Impacts are felt differently, but they are felt. The restorative benefits of exercise cannot be overstated and are a natural way of combating the realities of depression and anxiety. Getting in a good weightlifting session will not fix everything, but it does help to blow off steam and has obvious physical benefits. A run through the neighborhood will not erase the memory of that terrible accident, but it does get the blood pumping and the endorphins flowing. That in turn may help your firefighter remember how blessed they are to live in a good neighborhood, are healthy enough to run, and have a peaceful home to enjoy. These may seem like small things but when stacked together they are powerful weapons to ward off the accumulation of bad images that can pile up and cause problems.

Our counsel is to figure out a way to get some exercise together. You may not be able to do this all the time but get some mutual stuff on the calendar. Whether it's a game like tennis, a walk in the park, or a hike in the hills, how cool to spend time together and get a workout in. Anne and I have found our conversation during walks through our neighborhood or kayaking on our lake to be particularly open and incisive. The blood pumping and endorphins flowing seems to open us up, and we enjoy each other's ideas. Not sure what will work for you, just that something will if you give it a chance and make it a priority.

The bottom line is that exercise is a must and is doable no matter your budget or schedule. Have fun working this out and enjoy the benefits of feeling happier, healthier, and ready to take on all the other stuff this book is about.

GENERAL REFERENCES

Gavin de Becker, *The Gift of Fear: And Other Survival Signals That Protect Us from Violence* (New York: Dell, 1998).

Louann Brizendine, *The Female Brain* (New York: Harmony, 2007).

"Lactic Acidosis and Exercise: What You Need to Know," WebMD, https://www.webmd.com/fitness-exercise/guide/exercise-and-lactic-acidosis#1.

Jaime Coffey Martinez, "Help! What to Do about High Cortisol Levels," *Healthy Living How To*, February 5, 2013, https://healthylivinghowto.com/what-to-do-about-high-cortisol/.

John J. Ratey with Eric Hagerman, *Spark: The Revolutionary New Science of Exercise and the Brain* (New York: Little, Brown and Company, 2013).

NOTES

1. John J. Ratey with Eric Hagerman, *Spark: The Revolutionary New Science of Exercise and the Brain* (New York: Little, Brown and Company, 2013).
2. The information in this section is derived from Jaime Coffey Martinez, "Help! What to Do about High Cortisol Levels," *Healthy Living How To*, February 5, 2013, https://healthylivinghowto.com/what-to-do-about-high-cortisol/.
3. World Health Organization, "Depression," fact sheet, February 2017, http://www.who.int/mediacentre/factsheets/fs369/en/.
4. National Institute of Mental Health, "Major Depression among Adults," National Institutes of Health, n.d., https://www.nimh.nih.gov/health/statistics/prevalence/major-depression-among-adults.shtml.
5. Department of Mental Health and Substance Dependence, *Gender Disparities in Mental Health* (Geneva: World Health Organization, n.d.), 5–6, 7, http://www.who.int/mental_health/media/en/242.pdf?ua=1.
6. National Institute of Mental Health, "Suicide," National Institutes of Health, n.d., https://www.nimh.nih.gov/health/statistics/suicide/index.shtml.
7. Wendy LiKamWa McIntosh et al., "Suicide Rates by Occupational Group – 17 States, 2012," *Morbidity and Mortality Weekly Report* 65, no. 25 (2016): 644, https://www.cdc.gov/mmwr/volumes/65/wr/pdfs/mm6525a1.pdf.
8. Ratey and Hagerman, *Spark*.
9. Ibid.

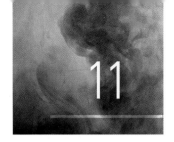

How to Sleep with a Firefighter

Most married couples who desire to stay married have to do one thing: get along with each other during the day. But for the firefighter couple, the task is twice as difficult as we have to get along both day *and* night. My firefighter works a 24-hour shift, and it is a blessing for many reasons. But the price we pay for these perks is some pretty serious sleep issues, as well as some quirky ones, that both he and I have had to learn to live with. It's the best job on earth and he's worth the extra effort, no question, but in order to survive his peculiar nocturnal habits, I've had to adapt. Hopefully the following tips will help other spouses out there who are struggling a bit to sleep with their firefighter.

Sleep deprivation is one of the most prevalent and pervasive issues that firefighters face. Most firefighters work a 24-hour shift, as described in chapter 3. At a certain point in the shift, say around 10:00 p.m., they may retire to their bunks and try to get some sleep. In many stations, when the alarm triggers, a light comes on in the bunkroom along with a loud bell, thus visually and audibly rousing firefighters from sleep and jolting them into immediate action (fig. 11–1). The volunteer is summoned from home by a radio or pager, which has essentially the same effect. Because of their sensitivity to these triggers, firefighters have difficulty falling into restorative REM sleep while on shift or on call. The problem often spills into home life or days off, making it difficult for your firefighter to *ever* fall into deep sleep. This can result in sleep deprivation, or chronic lack of sleep. Sleep deprivation is serious and should not be ignored.

I cannot stress the seriousness of this issue enough. Sleep deprivation has been linked to mental health problems, cancer, common colds, depression, diabetes, obesity, and strokes. It can affect job performance, as it impairs reaction time, judgment, vision, short-term memory, information processing, motivation, and patience.[1] These negative behaviors can affect home life as well, making your firefighter struggle to engage in family activities. Most critically, sleep deprivation is the best way to predispose oneself to becoming a stress casualty.

Fig. 11–1. Mike at a night fire, called to immediate action out of a deep sleep. Courtesy of John Odegard.

Ric & Joanne Jorge, Palm Beach County FD (retired), Florida

Some of the tolls that were consistent in my life (at work and at home) from lack of sleep were poor decision-making, slowed reflex reactions, and emotional outbursts. I remember being irritable the following day even when I slept through the night at the firehouse. My children were always glad to see Daddy come home, climbing all over me, loving on their very own hero. In my head I just wanted quiet and seclusion. What the hell happened to me? To enjoy my family and the life I had built, I needed to learn how to reset. This involved meditation and relaxation techniques. Including my wife in this created unity and more "us" time. It added greater value to our relationship because she knew she was a part of the solution I needed, which was something she would benefit from as well.

Firefighters compartmentalize tasks to manage and mitigate problems. This works so well that eventually we try to compartmentalize things like sleep, family, joy, love, and the like, instead of allowing these things to be spontaneous. In a career where we make order out of chaos it is very easy to fall into this mental trap. Let's call that mentality "hypervigilance." Hypervigilance is that mindset that races while you're in bed or left alone

> with your thoughts. Those thoughts that have you sitting with your back against a wall in a restaurant. Maybe you catch yourself doing a primary size-up looking for secondary means of escape as you are being shown your table. That hypervigilant/ruminative mindset also keeps you from having good sleep.
>
> Good sleep allows the mind and body to regenerate and heal itself, but for this to occur deep sleep must be achieved. Deep sleep is where rapid and nonrapid eye movement (REM and NREM) occur. Instead of having deep sleep, I slept light, so my stress level seldom went down and my brain and body never fully recovered from the rigors of shift work. The previously explained side effects were my result.
>
> To get that good sleep you must learn to control the thought processes of the mind. This will allow for more appropriate emotional responses and in turn lead to a different biological response (as described with the brain and body's healing in deep sleep). The person who develops this skill set will show up completely different. Your emotional spontaneity will return and your appreciation for life's most wonderful blessing, your family, will be fully realized.

Continual circadian rhythm disruption can lead to an absolute mental and physical breakdown. It is not meant to be constantly altered. Grossman explains how scientists first discovered the effects of circadian rhythm disruption, or sleep deprivation, during World War I (WWI). It was during this war that battles, for the first time in the history of warfare, raged for 24-hour, nonstop periods. Until then, soldiers had stopped fighting at night. In this new age of weaponry, soldiers could now see in the dark, and stress casualties were documented and labeled. In World War II (WWII), on the beaches of Normandy, soldiers experienced continuous fighting for 60 days and nights, and 98% of those soldiers became psychiatric casualties. In total, during WWII American forces lost 504,000 men to psychiatric collapse, due primarily to sleep deprivation.[2]

Research shows that you can die faster from lack of sleep than from lack of food. A study was done on rats and sleep deprivation at the University of Chicago in 1983. Rats kept awake became sick and died after just two and a half weeks. Those who became sick but were allowed to sleep made a full recovery.[3] Even fish need to sleep. Have you ever forgotten to feed your fish for a while? Did they die? Probably not. Fish die more easily from overfeeding than underfeeding. Leave the light on, however, and your fish will die very quickly.[4]

Research on long-term human sleep deprivation is rare because it is so dangerous to keep subjects awake for more than 24 hours. One study relates that after a night without sleep, drivers' ability to maintain their lanes was on par with that observed in drivers with a blood alcohol content (BAC) of 0.07%, and another study found that "deficits in performance accuracy and reaction time" in truck drivers after 28 hours without sleep were equivalent to those found in individuals with a BAC of 0.1% (legally drunk in the United States).[5] Essentially, "as continuous daytime waking exceeds 16 hours, psychomotor performance deficits increase to levels equivalent to BACs between 0.05 and 0.1%."

During sleep, your body is allowed to repair itself and your brain is allowed to process the day's events. The body secretes hormones while sleeping that affect growth and metabolic and endocrine functions, allowing the body to better operate the next day. Simultaneously, the brain is given time to store, process, and reorganize information, allowing your psyche to function at full capacity the next day as well. Sleep plays a major role in helping the brain with memory levels, retention, and the ability to learn new things. Have you ever noticed that you can go to bed after studying something and you wake up in the morning knowing the material? That's because during sleep, your brain is allowed to program the new data into your memory, just as if it were a computer. Without sleep, the brain is too busy functioning with wakefulness to process anything new into storage. That is how mental collapse occurs from sleep deprivation: your brain simply becomes too overloaded with incoming data to function properly.[6]

Enter in sleep, sweet, restorative sleep, "sleep that knits up the ravell'd sleeve of care."[7] Allowing the body a minimum of 4 to 5 hours of uninterrupted core sleep will absolutely work wonders on the psyche, with 7 to 9 hours being optimal. For the shift firefighter, a 30-minute nap can be effective, and a 2-hour nap can be highly restorative. We spouses must help our firefighters nap during the day when possible after a 24-hour shift and to sleep more soundly through the night by providing optimal sleep conditions.

TIPS FOR SLEEPING WITH FIREFIGHTERS

I've watched my husband Mike deal with this for more than 30 years now, and I've learned, over time, how to deal with it myself. Getting Mike to sleep has been, and continues to be, one of our greatest challenges. My babies were better sleepers than he is! Out of desperation, I've picked up ways to help him (and thus myself) and it is my hope that these tips may benefit some of you, as I believe that sleep is vital for good health, good performance, and good relationships.

Firefighters are just as tough to get to sleep as the average 2-year-old, so be patient

Why are small children so difficult when it comes time to sleep? Two words—they're wired and they're tired. They're wired because their little minds are learning and absorbing at lightning speed, and they're tired because their little bodies are still growing. It takes a very strong parent to help them relax without losing patience.

How is a firefighter like a two-year-old? They, too, are wired and tired because of the fight-or-flight response. Exposure to danger, as we all know, triggers a stress response that causes the nervous system to go into a state of high alert. For firefighters, this can happen every time the station bells ring, day or night. Firefighters become conditioned to be ready for action at a moment's notice, and this is called *hypervigilance*. Rick Jorge, in his video *Training Resiliency and Hypervigilance* from the excellent *Training Minutes* video series at FireEngineering.com, describes this as "an alertness that is necessary for prime decision making. Firefighters must 'turn it on' to do their job, but they get into trouble when they can't 'turn it off at home.'"[8] The senses are *heightened* for quick responses, making the firefighter oversensitive to light and sound and other stimuli, constantly scanning the environment for potential threats. Hypervigilance is necessary for survival in dangerous situations, but it comes at a great price as it is extremely exhausting. Thus, the firefighter is both wired and tired at the same time. The brain and the body conspire against the firefighter, impeding sleep.

As a firefighter spouse, I have to remember this when I find myself becoming irritated with my grouchy husband who obviously needs to sleep, but can't. With the patience of Job, I offer hot milk or tea, try to keep the house quiet, and encourage warm baths as I try to get him to relax.

Offer warm milk

Did your mother ever suggest that you drink warm milk to help you sleep? It turns out Mom was right: milk contains the amino acid tryptophan and the sugars needed to absorb it. Tryptophan is a natural sleep aid and pain reliever as well. And why is warm milk more effective than cold milk? It is believed to be because the few degrees that this drink may raise your body temperature are enough to trigger the slow-down response we experience on warm days. There is also anecdotal evidence that warm milk reminds us of infancy, hence we "sleep like a baby." I prepare a warm mug of milk that my firefighter thoroughly enjoys: simply microwave a cup of milk and add a splash of real vanilla, a dash of salt, and a teaspoon of sugar or artificial sweetener. It is both tasty and soothing and lacks the caffeine that is in hot cocoa.

Turn off the phone

Any noise easily awakens my sensitive firefighter. The last thing I want after all my efforts to help my firefighter get caught up on his sleep is to have him awakened early by an unnecessary phone call! Our phone rings off the hook all day every day, so on Mike's mornings to sleep in, I turn off the ringer and let the answering machine get it. Sometimes I forget to turn it back on, but oh well, a few missed calls is the price we sometimes pay!

Forgo those thin, frilly curtains

Again, because firefighters are hypervigilant (the busier the station, the more heightened the senses), they are sensitive to light. Light means action for the firefighter, thus any light in the bedroom will cause them to be alert. Everyone wakens with light, (this is nature's alarm clock,) but the firefighter especially so. Therefore, we have to fool Mother Nature by turning day into night. I've invested in heavy, light-blocking curtains that make our bedroom a veritable tomb. This makes the room a bit dark and dreary, but it's worth it when my firefighter is able to actually relax enough to sleep.

We have a dear friend in South Carolina who illustrates this problem in an amusing yet frightening way. He enjoys scuba diving in the alligator-infested, creepy critter–filled muddy waters of the Carolina rivers. He once told us that it is so dark and so quiet under those waters that when he gets down there in the black mud (to look for fossils and whatnot) he actually *falls asleep*! Underwater, with alligators no less! It's funny, but also somewhat tragic, that this light-sensitive firefighter is so sleep-deprived that when placed in total darkness, he falls instantly asleep. The threats of attack or even of drowning do not outweigh his desperate need for sleep.

Use a red-light night-light

As I've mentioned, turning on a bright light instantly arouses the sleeping firefighter to full wakefulness. I've accidentally turned on the overhead light in our bedroom before, and Mike will be halfway across the room before he realizes he's at home. So do yourselves a favor: use a dim, red-light night-light in your bathroom for those late bathroom trips or no night-light at all—a stubbed toe is better than rousing the sleepless one! These red lights are also being used in the firehouse bunkroom to lessen the impact of bright lights suddenly coming on.

Be a bit bossy sometimes

Tell your firefighter to avoid heavy meals before bed, at least two hours out,[9] though you may get resistance on this one as stress makes many people want to eat more, and trying to sleep can be stressful for the firefighter. When the body sleeps deeply, digestion slows way down. In fact, deep sleep is somewhat similar to being in a coma. If you have a large quantity of food in your gut, the body has to digest it, so it continues working instead of resting, which prevents full immersion into REM and deep sleep.[10]

Encourage your firefighter to exercise after a shift, though they may resist you on this one as well. Exercise is the only way to burn off any excess adrenaline that may still be in the system, as was laid out in much more detail in chapter 10. Adrenaline is a massive sleep inhibitor.

Don't let your firefighter consume caffeine or nicotine for at least six hours before bed, as both are stimulants. They will definitely resist you on this one. And discourage the use of alcohol as a sleep aid: although it is a depressant and may help them fall asleep, the subsequent metabolism that is needed to clear alcohol from the body can cause dehydration. This causes awakenings and is often associated with nightmares and sweats, so upon waking, they are likely to feel more sick than rested.[11]

A five-minute foot massage can quickly and easily help your firefighter relax and fall asleep. This one they won't resist. Look for more information on massage in chapter 13.

Tolerate the strange sounds in the night

Because of nightly disruptions at the station, most firefighters have a tough time reaching the deepest restorative sleep stage, where the body reaches a type of paralysis. A sleep study doctor told my husband this when his tests revealed that he'd only reached this stage for one minute, which, he said, he'd seen in other firefighters. People who stay in the lighter stages tend to talk in their sleep (this is called somniloquy).[12] My firefighter talks a lot. He barks orders, argues, and makes derogatory remarks. I am no longer alarmed by this as he is not directing them at me. He also laughs. Laughter is common amongst firefighters and is quite charming most of the time, but in the stillness of the night, it is rather creepy. So be aware of this side effect of shift work and don't be afraid: your firefighter is not insane, they just don't sleep as deeply as most people do!

You may also have to learn to sleep with other noises as well. A repetitive white noise, such as a fan, has been reported to help reach deeper sleep.[13] A specific sound will also let your firefighter know that they are home, allowing their subconscious to slip more deeply into restorative REM sleep. I've gotten used to the fan—it drowns out barking neighbor dogs and other disturbances as well. We now both sleep better with one than without.

Understand that they may fall asleep while you're talking

This is a good thing; do not be offended. We have found that one of the most helpful tips for inducing sleep is to recite the mundane, as it distracts a worried mind (e.g., counting sheep). Nothing puts one to sleep more quickly than a monotone, uninspiring classroom lecture, so view it as such. Be very careful about bringing up topics that will agitate or get their emotions going. I deliberately try to be as boring as possible—the more I ramble about unnecessary details, the sooner my firefighter falls asleep. For this, I'll talk all night—his sleep is even more precious to me than it is to him.

Don't be disturbed by the incoherent scribblings you may find on the bedside table

This is especially true if your firefighter is in charge of a busy firehouse. Don't be afraid, these are not signs of delusional schizophrenia but of a mind laden with heavy responsibility. Research shows that the number-one cause of insomnia in this country for any profession is worry or stress, including worry over to-do lists.[14] As people try to sleep, they mistakenly go over all that must be done the following day. This is the last thing any of us should do, as it automatically triggers a stress response that impedes sleep. The greater the importance, the bigger the stress response, and no work is more essential than a firefighter's. Lives are at stake; details must not be forgotten. We keep a pad of paper by the bed explicitly for this purpose because sometimes things come to my firefighter just as he's beginning to fall asleep. It is less disruptive to stay in bed and jot it down than to get up and do so. Once the idea is safely noted, he can let it go and drift off to sleep.

I know it seems a lot to ask a spouse to put up with, all this fuss over sleep. I'd encourage you to remember, however, that sleep is a wonderful gift to give your marriage. It restores your firefighter to their wonderful, patient self and motivates them to be more involved in family life. And let's face it, sleep will make them live longer and be more pleasant to live *with*. I think Shakespeare got this one right: "O sleep, O gentle sleep, Nature's soft nurse."[15]

Chapter 11: How to Sleep with a Firefighter

Sleeping with a firefighter is challenging business and not for the faint of heart. But hang in there, firefighter spouses—I have been told that firefighters *do* eventually sleep like normal people...about three months *after* they retire.

PITY THE FOOL

I do pity you poor spouses who love your firefighter and yet must figure out a way to navigate the choppy waters of sleeping with them. For most of us, it ain't easy. As Anne has so clearly described, lack of sleep is brutal. Both in the short term and long, the problems are significant and really impact your relationship. Let's face it, sleep deprivation is used as a very effective form of torture for a reason: it works. We need to minimize the instances and results of sleep deprivation or expect a torturous time trying to make home a happy place.

I am blessed to have a spouse who has made figuring out my sleep issues a personal crusade. She recognized early on that my lack of sleep was causing problems in our home and sought ways to help instead of just blaming me or getting mad. That seems to be a typical response for many of the couples we've spent time with and it simply doesn't help. Getting mad at the symptoms or repercussions does little to solve the underlying issue and usually leads to much bigger problems.

TIME IS NOT ON MY SIDE

We've talked about how great the 24-hour work schedule is with the flexibility it can provide, but here's a word of caution: the schedule can feed the sleep deprivation problem if you're not careful. On days off, it is very easy to stay up late, go off on adventures that are fun but not restful, travel to different places (and absorb the time-change fatigue), and simply overcommit. These activities are great in doses, but can end up adding to the level of fatigue and accumulating sleep deficit. Most of us can handle this on occasion, but when it becomes a lifestyle, you are asking for real trouble.

The reason this accumulation is bad for firefighters is what will be expected of them when they show up exhausted for work. The citizens don't know and don't care whether you're rested. They simply call with the worst problems of their lives and need answers. They need firefighters thinking clearly and able to provide help. And it's very important to remember that you'll be at their beck and call for 24 hours or longer. We don't show up to work and just need to get through an 8-hour day before we can go home and collapse from exhaustion. When we show up to work, we typically face at least 24 hours of chaos that require us to be at our best. Most alarming of all, fatigue and sleep deficit greatly compound the dangers that already exist in the work environment, putting you and your crew at greater risk.

For the firefighter, there must be a commitment to doing whatever it takes to get some quality sleep. I am preaching to myself here, as it has been one of the biggest struggles of my life. It requires giving things up that you love, sacrificing stuff that also makes life great, like late-night movies, social gatherings, sports programs the night before work, and projects that beckon. Not easy decisions and the tendency is to feel you can suck it up and make it work. That's usually true—until it's not. It's true until you show up to work drained and then must make critical life-or-death decisions with a brain that is not firing on all cylinders. It's true until you drag your fatigued carcass into work only to get called to that all-day fire that would have been a challenge even if you'd been fully rested. It's true until your body is pushed so far that the lack of restorative sleep caused a physiological breakdown that keeps you from doing the job you love so much. And it's true until the constant fatigue and the negative behaviors that result wear down your spouse until they can't take it anymore. On that day, you'll wish you'd planned for, sacrificed for, and succeeded in getting the rest you needed.

NOTES

1. Michael J. Breus, "Sleep Habits: More Important Than You Think," WebMD (2004), https://www.webmd.com/sleep-disorders/features/important-sleep-habits#1.
2. Grossman and Christensen, *On Combat*.
3. C. A. Everson, B. M. Bergmann, and A. Rechtschaffen, "Sleep Deprivation in the Rat: III. Total Sleep Deprivation," *Sleep*, February 1989 12(1): 13–21, https://www.ncbi.nlm.nih.gov/pubmed/2928622.
4. Allan Rechtschaffen, *The Psychology and Physiology of Sleep*, University of Chicago, National Institutes of Health, http://grantome.com/grant/NIH/K05-MH018428-26.
5. Jeffrey S. Durmer and David F. Dinges, "Neurocognitive Consequences of Sleep Deprivation," *Seminars in Neurology* 25, no. 1 (2005), https://doi.org/10.1055/s-2005-867080.
6. "Brain Basics: Understanding Sleep" (Bethesda, MD: National Institute of Neurological Disorders and Stroke), https://www.ninds.nih.gov/Disorders/Patient-Caregiver-Education/Understanding-Sleep.
7. William Shakespeare, *Macbeth*, 2.2.47 (Moby Shakespeare, 1995), http://shakespeare.mit.edu/macbeth/macbeth.2.2.html. References are to act, scene, and line.
8. Ric Jorge, *Training Minutes: Firefighter Resiliency and Hypervigilance*, FireEngineering.com (January 24, 2017), http://www.fireengineering.com/articles/2017/01/training-minutes-firefighter-resiliency.html.

9. NIOSH, "Plain Language about Shiftwork."
10. Mark Reddick, "Should You Eat Just before Bed?—All the Risks Explained," The Sleep Advisor (April 12, 2017), https://www.sleepadvisor.org/eat-just-before-bed-risks/.
11. "Missouri University Study: Alcohol Is Not a Sleep Aid," OzarksFirst.com (December 26, 2016), http://www.ozarksfirst.com/news/missouri-university-study-alcohol-is-not-a-sleep-aid/631694789.
12. "Sleep Talking—Causes," National Sleep Foundation, https://sleepfoundation.org/sleep-disorders-problems/sleep-talking/causes.
13. Meghan Neal, "The Many Colors of Sound," *Atlantic*, February 16, 2016, https://www.theatlantic.com/science/archive/2016/02/white-noise-sound-colors/462972/.
14. Ibid.
15. William Shakespeare, *Henry IV*, part 2, 3.1.5–6 (Moby Shakespeare, 1995), http://shakespeare.mit.edu/2henryiv/2henryiv.3.1.html.

WHEN A FIREFIGHTER SPOUSE SLEEPS ALONE

This is something I've never really talked about or told anyone outside my immediate family, as I am quite embarrassed by it. I've worried that if people knew about it they'd think I was nuttier than an outhouse rat. To be honest, I've secretly feared that maybe I am. Since I've never discussed it, I don't know if other firefighter wives struggle with it or not; so I'm taking a chance, baring my soul, and exposing a weakness with the intent of perhaps helping or encouraging another firefighter spouse out there who may share this problem and is, as I have been, keeping it to herself.

Everyone knows that firefighters can develop sleep issues, but so can their wives. I'm keeping this to wives, as I don't think many husbands struggle with this problem because it stems from a fear of being alone. At night. For many women, sleeping alone sometimes becomes a challenge with frightening results. Ever since my husband Mike became a firefighter more than 30 years ago and began working 24-hour shifts, I've struggled on and off through the years with a phenomenon that is, for me, a direct result of being afraid and alone—*night terrors*.

Night Terrors, the Story

A night terror, or *pavor nocturnus*, is *not* a nightmare. It is far, far worse. How did mine begin? I'll tell you.

It's 1986, and I am a young wife and mother living in an apartment with my husband and newborn son, Michael, in Pineville (LA). I am far from home, far from family and friends. Mike is stationed at England Air Force Base where he is serving as a crash firefighter in the United States Air Force (fig. 12–1). We've been in Louisiana about six months now. With this new career, Mike works 24 on, 24 off, then every two weeks he has a "Kelly day," or day off, giving him a three-day break. For the first time in our marriage I must get used to "sleeping single in a double bed" every other night, which gets a little confusing. We've just moved to this apartment from our old apartment across the river in Alexandria. We moved here because a woman was raped at the old place and I could no longer bear to live there, for that was my worst fear. This new place is much nicer and I'm beginning to feel safe again.

Fig. 12–1. Mike with his crash firefighter crews in the United States Air Force (bottom row, far right)

Then one day I get a crank call, a "sex call," while Mike is on shift. At first, I think it *is* Mike, because it sounds just like him, so I laughingly go along with it. But I soon realize it's *not* him, and hang up in horror. Mike and I later chuckle about it, but the problem is, the guy keeps calling back, and he happens to call *every day* that Mike is on shift! I begin to fear that this person is watching me, that he maybe lives nearby and can see when Mike's car is gone. Or perhaps he's another firefighter who knows his schedule. My anxiety mounts as day after day the stranger calls.

One night I go to bed, alone and afraid, for I'd had another call that day. Our room is never quite dark because of the streetlight outside. I hear something. I open my eyes—a man is standing by my bed! He reaches for me. I gasp in surprise and begin to scuttle away from him. I blink and blink, trying to see in the dim light. He's there, I can't believe what I'm seeing, a stranger is in my house threatening me! I continue to blink, trying desperately to get a good look at him. I turn on the light—there's no one there. My heart is pounding with a rush of adrenaline that courses through my body. I'm breathing rapidly, even sweating now, I am in full fight-or-flight mode. I cannot even begin to describe the intensity of my fear, for the person standing there was as real as real can be, if only for a moment. My subconscious mind, in the early pre-REM stage of sleep, played a horrible trick on me and my body couldn't help but respond. I've just had my first night terror.

A few nights later, I again open my eyes to see a figure standing by the bed. I scream, I begin to scramble away—someone grabs me. I fight, I kick, I am after all, alone, so who's touching me? Then a light turns on, it's my husband. He's frantic, asking, "What's wrong?" He, too, is in full fight-or-flight, as his wife just started screaming. In my subconscious state, Mike had become part of my night terror. His erratic and irregular presence in my bed confused my sleepy mind. I apparently have a problem and my problem has now become Mike's problem too.

I'm horrified, embarrassed, and ashamed that my anxieties over being alone at night have affected my husband's sleep. He's already sleep-deprived from the firefighter schedule, and now his sleep is being disrupted at home, thanks to me! My only desire is to help my husband, instead I am adding to his burdens.

We both intuitively knew that this was a result of his new career combined with my phone stalker, as I'd never had these night terrors before, but what were we to do? Mike had to work a 24-hour shift, so I had to be OK with that! Fortunately, the phone calls ended: one day he called when Mike was home. He'd been scheduled to work that day but he made a last-minute trade. The calls always began with the guy saying, "How'd you like me to…[insert something dirty here]." That day, Mike grabbed the phone and angrily yelled, "How'd you like me to kick your @*&?!!" That was the last crank call. I began to hope that my anxiety would diminish, putting an end to the night terrors. But they didn't end.

Mike put wooden bars in the windows and sliding door for added security. We bought a gun and he taught me to shoot. We got involved in a great local church and made friends. These things helped, and the terrors decreased from nightly down to weekly, but they continued, so it wasn't over. It was the schedule, it was sleeping alone, and that wasn't going to change. I, as a young wife, began to fear that I had a defect, a flaw that rendered me somewhat inadequate to be a firefighter's wife. I was afraid to sleep alone at night and this fear was manifesting itself in a horrific and disruptive way. My night terrors continued regularly and grew in intensity until the spring of 1987, when a climactic event occurred that, oddly, resulted in a reprieve.

It was a night in March, and we headed to bed. Mike was particularly fatigued; he was taking a life-saving course that required massive amounts of swimming. He swam miles a day as he trained to become certified in water rescue, something he felt would add to his firefighting resume. His shoulder muscles were tight, *very* tight from the exertion. He drifted off to sleep easily.

I, on the other hand, struggled to fall asleep. My recurring night terrors over the past seven months or so had turned bedtime into a daunting and frightening event. I was trapped in a vicious cycle of unrelenting anxiety—anxiety over sleeping alone and anxiety over sleeping at all, for the horror of waking up to someone standing by my bed was too terrifying for words. I kept wondering, "When I close my eyes, will *he* come tonight?"

I sensed Mike's concern and frustration at the utter helplessness of the situation. He wanted to help, but couldn't. He wanted to work a day job to be home with me at night, but that was not available. He wanted them to just go away, but they wouldn't.

I was so ashamed and frustrated with my night terrors, but I didn't know how to make them stop. They weren't every night, but nearly. I never knew when I'd have one—it didn't seem to matter anymore if Mike was there. I began to despair, to believe there was something really wrong with me. I prayed. I cried out to God to make them stop. I was so very tired of living in fear. It was beginning to cripple me. In a roundabout, seemingly awful way, God did answer my prayer.

That March night, I finally began to drift off to sleep. Then he was there, leaning over in the darkness, reaching for me—the faceless man who terrorized my sleep. I sat up, screamed, and scrambled away. Mike rolled over and quickly moved to confront whoever was in our room. I was fully awake now, Mike was comforting me. I began to cry with frustration, shame, and fear. "It's okay," Mike said in the dark, patting me once again. It had become routine. To his credit, my husband has never once said anything cruel or derogatory about my night terrors.

"I'm so sorry," I sobbed, wishing with all my heart that I could quit disturbing his sleep. It was burdensome enough to disrupt my own, but to do it to him as well was simply unbearable.

I began to calm down; my pounding heart began to quiet. "Go back to sleep, it's okay," Mike said again. I lay back down. Mike remained sitting up. "My shoulder hurts," he said. "It's really starting to hurt."

I got up and turned on the light. I looked at Mike sitting there. Something was askew. I came closer and we both looked at his right shoulder—it wasn't there! To our utter surprise, the arm was about 4 in. lower than it was supposed to be—it was out of the socket! He must have been laying on it as he bolted upright after hearing me scream. This, combined with the muscle tightness from his swim training, was apparently enough force to dislocate his shoulder. My problem had now actually caused physical injury to my poor husband!

Mike began to moan and writhe in agony as the pain hit him full force. We had to get him to the hospital, and fast! I helped him up, still in shock at what

I had caused, and left him to try to throw on some clothes. I ran into baby Michael's room and snatched him up. Mike could barely stand, the agony was so overwhelming. I loaded him and the baby into the car. Thankfully, 10-month-old Michael sat quietly in his car seat, wide-eyed and bewildered by this late-night adventure. He'd never heard his dad moaning and carrying on in such a fashion.

We fought a bit on our way to the base hospital, as I only knew one way to get there and that way was blocked by the longest train in the history of trains. "Go the other way!" Mike yelled. "I can't wait any longer!" Now what? I didn't know the other way and he was in no shape to tell me. Eventually, somehow, I got us there. It truly is hard to think when in full fight-or-flight mode.

The hospital staff had never heard of anyone dislocating their shoulder in bed before. Needless to say, we got several looks and heard lots of snickers and suppressed laughter. Mike and I would be the butt of many jokes on base for months to come.

As a result of Mike's injury he was put on light duty for about eight weeks. He worked days and was home at night. This reprieve broke my cycle of night terrors. His being home regularly helped me to completely calm my anxieties, which had been compounding and spiraling out of control. And during that eight weeks we found out I was pregnant with baby number two—Ricky. The jokes about that fateful night are now the stuff of legend!

With this new pregnancy, my fatigue level caused me to fall asleep very quickly. Even when Mike returned to shift, my night terrors did not. I rejoiced! Then once I had Rick, I was so tired from caring for a newborn while chasing a toddler that I continued to sleep easily. I did not have a single night terror in Louisiana after the "shoulder incident," as it came to be known.

But unfortunately, my saga does not end there. The night terrors returned some years later. I've been forced to learn much on this topic, as it seems to be an ongoing issue.

NIGHT TERRORS, THE DEFINITION

A night terror is a parasomnia disorder, or sleep disorder, in which a person quickly awakens from sleep in a terrified state. During a night terror, the person sees things that aren't really there. Night terrors occur within the first few hours after falling asleep, during stages 3 and 4 of non-rapid eye movement (non-REM) sleep.[1] Sleepwalking (and bed-wetting) occur in this same stage. It is a mystery why people have night terrors, as during this stage the brain is supposed to be completely shut down. *Nightmares* occur in the dream state of sleep, or REM, and usually happen during the second half of the sleep period.[2] Nightmares do not cause the same reactions as night terrors because the body recognizes that they are not real.[3]

It is suspected that night terrors can be passed on genetically.[4] This is interesting to me, as my dad was a sleepwalker as a child (so was I), and he, too, suffered from night terrors as an adult. Chronic fatigue, emotional fatigue, anxiety, and mood disorders are common in adults experiencing night terrors.[5] My father was an army intelligence officer, or spy, who monitored communist activity in East Berlin in the 1950s. His life was threatened on numerous occasions and he was simultaneously traumatized by the deaths of many of his comrades. My mom told me that during the early years of their marriage she'd awaken to him screaming, grabbing her by the hair, or wandering around the room patting the walls as if looking for an escape. Post-traumatic stress disorder sufferers commonly report having night terrors. They generally involve either a shadowy figure threatening them or having bugs or snakes in the bed. I've had all three (yes, I've actually felt like there were bugs and snakes crawling on me under the covers), but mine have predominantly been the shadowy, faceless-figure kind, as I seem to fear men more than creatures. The terrors I've experienced through the years are not from the stress and trauma of risking my life, but from the anxiety first triggered by a sexually explicit phone stalker (who seemed to know my husband's schedule) and years of sleeping alone.

As I have already detailed, night terrors are *very real* to the one experiencing them. They can cause a temporary inability to regain full consciousness, thus triggering a full fight-or-flight response, as you cannot discern reality from fantasy. Heart rates have been measured to spike as high as 160–170 beats per minute, which is astoundingly high for someone who was, by all appearances, just resting.[6] This is because of a sudden dramatic surge of adrenaline, something firefighters can also experience when awakened at the fire station. I have felt actual pain as my heart thuds so hard. It is the body's attempt to prepare for *immediate* action, to go from zero to full speed ahead in an emergency—whether real or perceived.

NIGHT TERRORS, THE SOLUTION (SORT OF)

It is best not to touch a person who is having a night terror, for this can cause further alarm and confusion (as I have experienced). People have been documented committing *involuntary manslaughter* in this state; so deeply are they immersed in their terror that even a loved one can be mistaken for an attacker.

So do not try to touch or shake them awake. Instead, gently talk them out of it; this tends to be enough. In extreme cases, it is best to just leave the room and lock the door for your safety and theirs. And in these extreme cases, it is also recommended to move hazards so the person does not trip over them.

Eating a heavy meal before bed prevents deep, REM sleep and can cause a night terror.[7] So does coffee, and I know this from personal experience. Chronic lack of sleep has been correlated to night terrors. Reading, a hot bath, warm milk, and foot massage all instigate sleep quickly and easily. These things have helped both my husband and me with our sleep issues.

You may wonder if I still have night terrors, even after all I've learned about them. The answer is yes, unfortunately, I occasionally still do. They come and go. They stopped when I lived in Louisiana, but then they resumed for a bit when Mike first became a Seattle firefighter. But this time they settled down quickly, perhaps because I was back in my hometown, which was familiar territory, or perhaps because my boys were a bit older—I'm not exactly sure why.

I've discovered that they can occur in a new house or even at a hotel—situations that are unfamiliar and inhibit sleep. Through my boys' teen years, I didn't have a single one; I guess they made me feel safe. When they first left for college, Mike had simultaneously made captain and been assigned to an administrative position at the JTF, which rendered him home nights for two and a half years. Though most firefighters hate being off shift, as mine certainly did, I secretly rejoiced as I feared that my boys being gone might cause my terrors to return.

I was right to be afraid, because the terrors made another, sudden appearance. Mike began teaching air management, which involved traveling. I was once again alone at night and for even longer periods than just 24 hours. My precious boys began taking turns coming home from college (just an hour away; fig. 12–2) to "babysit" Mom for the weekend when they could. They knew their presence would help me overcome the terrors before they got too severe. Yes, this is a family concern, and all three of them care about my fears. The terrors have ceased for now, but Mike has long since returned to shift and his teaching career has only increased. I am alone at night more than ever before and very aware of the fact that the terrors may suddenly return—at any time.

Over the years, I have become more afraid of the night terrors than of any real threat—they are that traumatic. Mike has considered going off shift because of my problem. He is so completely devoted to my welfare that he is willing to sacrifice his beloved station life for a desk job, and makes repeated offers to quit teaching. But I won't hear of it.

Fig. 12–2. Our college-age boys visiting Dad at the JTF. Left to right: Rick, Mike, Michael.

In my efforts to combat night terrors, I now avoid drinking coffee too late, watching scary movies alone, and eating heavy meals before bed. It's silly, but locking my bedroom door helps a lot. So does prayer. I moved the family dog kennel into the bedroom, something I swore I'd never do, but the dog's presence was a huge comfort and worth the hair, smell, and incessant licking (the dog, not Mike.) Since anxiety seems to be the primary cause, finding ways to alleviate it is a solution. Exercise helps too as it burns off excess adrenaline and aids in sleep—for us both.

On the rare occasion that I have another night terror now, thankfully I no longer scream, I simply gasp or whine. I sometimes still scuttle away from the stranger by my bed, but not always. Mike and I have learned to pretty much ignore the terrors: I don't react as strongly as I once did, and neither does he. I usually don't even wake him anymore, thank goodness. But there is still that brief moment between wakefulness and semiconsciousness when I have to ask myself, "What if this time there really *is* someone there?" This fleeting thought is enough to make me react, if only just a little.

I hope my story helps a fellow sufferer a bit, even though I offer no absolute cure. Sleeping alone can be unsettling at times as it is a scary world—fear isn't irrational—just real. Have courage, firefighter spouse—you're not the only one who is, sometimes, afraid of those bumps in the night.

Chapter 12: When a Firefighter Spouse Sleeps Alone

Paul & Sheryl Combs, City of Bryan FD, Ohio

We'd like to tell you a story. Two college kids walked into a month-long winter-term class to experience a historical painting course from one of the area's most prolific painters and a Defiance College English professor. One was a cocky and self-assured artist who knew he was destined to be the next multibillionaire Norman Rockwell; it was only a matter of time until everybody else discovered what he already knew—he was awesome. The other was a hardworking freshman who was enjoying college independence and had life mapped out according to her religious and family beliefs. He saw her as a cute but wide-eyed freshman who brought no benefit to his inevitable fame-filled career. She thought he was a partier and an atheist. This is how our story began.

We became friends over that month-long class, but nothing more than friendly conversation ever materialized and we went our separate ways when the class ended. Fast forward eight months and life had dramatically changed for us both. For Paul, his mother's cancer diagnosis had forced him to drop out of college and get a job as a stockman at a local big-box store. During this time, he also discovered that art is hard work and the world was not going to knock on his door with fame and fortune in tow—this was a brutal epiphany. Sheryl broke up with her long-time boyfriend and also experienced family financial hardships that refocused her on work and helping her parents weather their financial burden. Her glossy-eyed view of the world had been sharpened. So, by the time we met again for the fall semester of college we were both drastically different people.

We came together during a shared literature class where our friendship was renewed. It was nothing more than friendly chatter, which quickly became intimate talks and long campus walks. Then came the first date, more dates, and finally the first kiss, which sealed the deal for us both.

We tell you this story not because it's special in any way—millions of people meet like this every year—but to stress the importance of friendship within your relationships. Over the past 25-plus years of marriage, it has been friendship that provided the rock-solid foundation for our successful marriage and faithful life together. Hard times have come and gone and challenges have strained our bond, but never our friendship. Like any well-built home, the structure is only as good as its foundation.

If we are to offer any advice to young couples just beginning their journey together, or the couple trying to regain their once brilliant spark, it is to focus on friendship first. Money, success, fame, family, children, not even intimacy can fix a relationship void of friendship. We are not marital experts—far from it. We struggle, we have our differences (we are polar opposites in almost everything), but after 25 years we have experienced what a strong friendship based on respect can do. It's like magic, with the rabbit in the hat.

We are first and foremost friends. We cherish this, and, like the cartoon drawn by us both suggests (another first), we build *all* on the foundation of friendship.

SCARY AND TOUGH

These were challenging periods for us. And then add the helplessness you feel over what happens while you're sleeping. Since I'm already the poster child for the lack-of-sleep all-star team, you can imagine how having your wife start screaming in the middle of the night works out.

I believe it is facing times like these together that gets to the heart of just about everything we are writing in this book. Believe me, it is frustrating and at times terrifying to go through this stuff. But we pursued it as a team. And as a team, you may have to do some things that are less than desirable for you, personally. It always comes back to what is best for us, not just for me.

IT'S ABOUT US

One of the things we had to consider (and then ultimately do) was giving up my beloved 24-hour shift. I don't feel you should do this recklessly, but many of my coworkers have taken stints in administrative jobs as a solution to things that are going on at home. Whether it's the night terrors like we faced or just a period that needs a more stable routine, sometimes a break really helps your spouse and kids.

An example of this is a friend whose wife was going through a rigorous academic process to finish off her master's degree. The more stable schedule fit much better, allowed for child care to be arranged, and gave his wife peace of mind that allowed her to excel in her studies. This type of arrangement is pure gold in our view and represents the sacrifices that really forge solid marital bonds. Working the day job was a drag. Not going to fires was terrible. Braving the arduous city commute was soul stealing. But what an incredible display of doing what's best for "us."

WATCH IT, PAL

Another key point to make, whatever the challenge, is to be very careful about your tongue. This is especially true when it is something personal like the night terrors. Anne was truly afraid, truly frustrated that they were happening, and a bit fragile emotionally that she couldn't help it. Now bring my reactions into the equation. I'm also tired, frustrated, and feeling helpless. This is a recipe for insensitivity, misunderstanding, and mockery that would make matters significantly worse.

To be more direct, firefighters, I'd encourage you to really watch how you joke about these types of things. Humor can be very helpful and lighten the mood in tense situations. But please recognize that we as firefighters are used to a very brusque, at times coarse, version of humor. That does not play here and typically will cause further wounds, deeper frustration, and a heightened sense of insecurity, all coming at a time when they really need you the most.

YOU'RE NEXT

The final recommendation is to be that supportive team you dreamed of being when you got married. To the firefighter going through these types of struggles with your spouse, I'd encourage you to remember that your time is coming. You will be injured or face depression even if you can't imagine it now. Most every firefighter gets to battle difficulties on a personal level that arise from doing our work. How you handle your spouse's challenging times will go a long way toward setting the rules for your turn at the plate, when and if it comes.

Hopefully, as we did with the night terrors, you'll face whatever challenges that come against you together, arm in arm, as a team.

NOTES

1. Shailesh Jain, "Sleep Terrors in Adults: How to Help Control This Potentially Dangerous Condition," *Current Psychiatry* 11, no. 9 (2012): E1–E2. http://www.mdedge.com/currentpsychiatry/article/64825/sleep-medicine/sleep-terrors-adults-how-help-control-potentially#.
2. Lizette Borreli, "Night Terrors in Adults: When Sleeping Turns to Terror After Dark," *Vitality* (blog), *Medical Daily*, February 6, 2015, http://www.medicaldaily.com/night-terrors-adults-when-sleeping-turns-terror-after-dark-321100.
3. "Night Terrors: Why They happen and What to Do about Them," Babycentre, October 2008, https://www.babycentre.co.uk/a1022133/night-terrors-why-they-happen-and-what-to-do-about-them.
4. Ibid.
5. Jain, "Sleep Terrors in Adults"; and Borreli, "Night Terrors in Adults."
6. David W. Richards, "More Information," Night Terrors Resource Center, n.d., http://www.nightterrors.org/more-information/.
7. Richard Foxx, "Fighting Fright: How Adults Can Manage Night Terrors," Doctors Health Press, November 3, 2015, https://www.doctorshealthpress.com/mental-health-articles/fighting-fright-how-adults-can-manage-night-terrors/.

The Healing Power of Touch

Talking about stress and trauma does help a lot of the time. But sometimes trauma is so deep that talking about it does not help, especially not for men, and especially when traumatized by near death experiences. In fact, sometimes talking can actually retraumatize. Erich Maria Remarque describes this struggle best in his novel *All Quiet on the Western Front*. As a soldier on reprieve from the war, he is constantly asked to describe his experiences. He keeps his stories humorous and lighthearted, even when family and friends press for the darker tales. He describes his discomfort to answer his friend's probing questions when he writes: "I realize he does not know that a man cannot talk of such things; I would do it willingly, but it is too dangerous for me to put these things into words. I am afraid they might then become gigantic and I be no longer able to master them."[1]

As a longtime firefighter spouse, I have recognized my husband's reluctance to speak of his tough runs (fig. 13–1). I wish to comfort him, to ease his anxiety, but often my attempts to elicit conversation have fallen flat. But what he *has* responded to more so than anything else from me is this: *my touch*. Research by Bessel van der Kolk, an expert on trauma, suggests that traumatic memories are often stored as nonverbal sensory, motor, and emotional fragments, so that no matter how much verbal therapy you do, you may never get to the core of it. What he and many others recommend is therapy that includes physical elements, such as exercise or yoga, or human touch, such as massage. It has been suggested that touch can ease stress, anxiety, and even pain on a physical level beyond what mere words can do.[2]

This may be one of the reasons why married firefighters have lower rates of PTSD than unmarried ones, because when it comes to being touched, they literally have the upper hand. Those who have supportive relationships recover more quickly from trauma than those who do not, and physical touch aids in that recovery. The loving ways couples touch have many benefits in combating stress.

CHALLENGES OF THE FIREFIGHTER MARRIAGE

Fig. 13–1. Sometimes words escape my firefighter. He carries a very heavy load. Courtesy of Lt. Tim Dungan, Seattle FD

Cuddling: Several studies have been done on the effects of hand-holding, hugging, and kissing in married couples and their levels of stress. In one such study, 16 happily married women were subjected to the threat of a mild electric shock and their anxiety levels were measured. Some were not touched, others held hands with a stranger, and the third group held hands with their spouse. The group without a hand to hold had the highest level of "anxiety or threat-related brain activity," the group holding a stranger's hand had slightly less, and the group holding the spouse's hand showed *significantly* lower levels of anxiety.[3]

The destressing effect of touch was also measured by a similar study published in *Behavioral Medicine*. Cohabitating couples either held hands for 10 minutes and then hugged for 20 seconds, or told not to touch at all before exposed to a stressful situation. The huggers had healthier reactions to stress than the nonhuggers, including lower heart rates, lower blood pressures and smaller heart rate increases.[4] The gentle pressure of a hug has been shown to stimulate nerve endings under the skin that send calming messages to the brain which slows the release of the stress hormone cortisol. A hug makes you feel safe.

The power of a kiss has been revealed in a study on 2,000 couples. Those who kissed daily were *eight times* less likely to struggle with the effects of stress than those who only kissed during lovemaking. Kissing is a close connection that is proven to release endorphins that directly counters the effects of cortisol.[5]

Massage: The benefits associated with massage are experienced from head to toe—it doesn't matter whether it's back, neck, or foot. The human touch is healing, soothing, and relaxing. The word "massage" is derived from the Greek word *massein*, which means "to knead, to handle, touch, or work with the hands." People fall asleep all the time on the table of a masseuse. Why is that? Massage has been shown to release feel-good hormones or endorphins into the body and brain. Endorphins directly reduce levels of norepinephrine and cortisol, which are stress hormones and primary inhibitors of sleep. Massage also decreases adrenocorticotropic hormone (the hormone that stimulates cortisol production) and nitric oxide (a free-radical gas).[6] A study was done on subjects who suffered from either depression or insomnia or both. The control group was given a daily massage over a period of five days. At the end of the five days, the control group's urinary levels of cortisol were much lower than the nonmassage group's with an average decreased level of 31%. The control group also experienced an increased number of hours slept at night and reported lowered symptoms of anxiety, with an average increase of 28% for serotonin and an average increase of 31% for dopamine.[7]

Endorphins released during massage have other benefits as well. Not only do they counter stress and lift the mood, but they also prevent nerve cells from releasing more pain signals to the brain, literally "making the pain go away." Some studies have suggested that massage may increase levels of beta-endorphins, which would decrease pain levels.[8] And one study even noticed amped-up levels of white blood cells, a vital player in the body's immune system.[9] A sore and weary firefighter cannot relax and go to sleep if pain impulses throb loudly through their brain. Pain keeps the body awake, and remember that the sleep-deprived firefighter is more prone to addiction, depression, and becoming a stress casualty.

Phil & Meg Arensdorf, Dubuque FD, Iowa

Regardless of what profession you marry into, there is no "easy" way to be married. The biggest lesson we have learned is that marriage is hard and takes continuous work. The workload varies in size, but it is always present and should be focused on. So many things affect it—work, money, kids, obligations to family and friends. Our takeaway has been that if we keep chipping away at it with dedicated and scheduled time, we're able to use the outside influences in our lives to make us stronger as a couple rather than allow them to separate us.

It wasn't always like that for us—it took a lot of pointless arguments and getting upset over the mundane for us to realize how much time and energy we were wasting on the negatives. We went through a rough patch of truly thinking we were going to split up about seven years into our marriage, and dedicated ourselves to working on it in intense couples counseling. That is where we learned the benefits of scheduled time. We spend two hours each week, just the two of us, face-to-face. No movies or TV shows, no computers, no phones. We play board games, do quiz books, go for walks. Sometimes our dates are at home, after our children are in bed for the night. Sometimes we go out—we've been known to grab a few beers together and just chat, go zip lining, paint pottery. We keep ourselves open to any adventure, whether it be an old favorite or something new for the both of us.

We schedule our time together usually one month at a time. On a Sunday evening, we'll compare calendars and pen each other in for each week. While we've had periods over the years where we haven't been as diligent, it is the number-one thing that works to put us back on course. Between Phil working 24-hour shifts at the fire department and his job with the college, scheduling time for each other has been a must. Otherwise it's too easy to keep pushing each other to the back burner because of other responsibilities. While we will be sad to have our boys leave, when we become empty-nesters, we want to see it as an exciting and door-opening experience. At that point in our lives, we will have unlimited access to each other and can make every day an adventure, should we choose.

We genuinely like each other and care about the other's interests. We not only support each other's passions and goals, but we actively participate in them—there's no being passive in marriage when it comes to leaning in to your spouse and helping them accomplish the next thing on their list of goals. Let love win in life, it's worth it!

THE FIVE-MINUTE FOOT MASSAGE

I've spent many years agonizing over my husband Mike's inability to sleep because of his firefighting career. Then one day a miracle occurred: Mike fell asleep quickly and easily, and without even trying! I gaped open-mouthed at this phenomenon—Mike, sleeping, with the lights on and everything. What had happened? What strange, magical fairy dust had just materialized to cause this event? I'll tell you: it wasn't fairy dust. It was, quite simply, a five-minute foot massage. We'd been sitting watching a movie together, a common evening activity for us, when I took his tired, aching feet onto my lap. I began to rub them half-heartedly and without much effort as I watched the show. I kneaded them like Christmas cookie dough, and the next thing I knew, he'd passed right out. Since that day I've become a firm believer in the power of foot massage.

The reason I prefer to give Mike a foot massage as opposed to say, a back massage, is that it's easiest for me while being just as effective for him. I'm not strong enough to give him a deep-tissue massage through his thick muscle layers. I try on occasion, with little effect. (He, however, has a great effect on my bad shoulder when he massages me, as I have little muscle and he is very strong.) But if I rub his feet, which have less muscle and thinner skin, I strike gold, as the feet are warehouses of stored stress.

Situated in each foot is the *solar plexus reflex*, a spot where much of the body's stress resides. When pressed, the solar plexus reflex can calm the entire nervous system. It is located in the center of the ball of the foot (near the toes.) There are approximately 7,000 nerve endings in the feet—when rubbed, the release of stress is felt throughout the entire body, including the brain, as these nerve endings are connected to every system. Foot massage can sometimes even relieve headaches.

But wait, it gets better. Foot massage improves circulation and heart health. Blood transports oxygen and nutrition to the body's cells. Blood also works to remove waste and toxins from the cells, cleansing your system. When stress is present, whether from real danger or merely perceived, blood flow becomes limited, as the body believes there's no time for things like cleansing and repair. It's all hands on deck, emergency mode, meaning blood is used for strength purposes only. For firefighters this is a problem as their job forces them to be amped-up on adrenaline, meaning there is never time for things like cleansing and repair. Stress hormones keep the heart working overtime, which can eventually lead to heart disease, a major killer of firefighters. Endorphins are the key to eliminating these stress hormones, which wreck such havoc on the circulatory system, and there are two easy ways to create them: one is through exercise, the other massage.[10]

Foot massage may prevent harmful effects of stress from taking a toll on other areas of the body as well. If present, stress hormones can upset the digestive system, muscular system, and skeletal system. The immune system, too, can be worn down by the effects of adrenaline. But the power of endorphins not only eliminates stress hormones, but it actually stimulates immune cells as well, making them stronger and better able to fight disease. These diseases can range from the common everyday cold all the way up to the more serious ones like cancer. A strengthened, invigorated immune system may even slow aging.[11]

Foot massage promotes sleep, lifts the mood, eases pain, and counters illnesses associated with stress. What a wonderful, simple gift to give to the one you love. The power of the human touch is pretty amazing stuff, isn't it?

An additional benefit is the release of the hormone oxytocin into the brain. Oxytocin is the hormone generally associated with orgasm—but a 5- to 15-minute massage raises oxytocin levels as well.[12] Oxytocin is the hormone of love; it triggers bonding, relaxation, peacefulness, and feelings of generosity. Because of oxytocin, a massage sets the stage for a romantic evening. It is sensual, enjoyable, nurturing, and loving. What a wonderful way to show your spouse how much you care!

Foot massage tips

Any type of massage triggers positive bonding results but the easiest one to perform at home is foot massage, therefore it is my massage of choice. Here are some tips to giving the perfect foot massage.

It is best to soak your feet in warm or hot water for at least 10 minutes prior to massage. Have your loved one take a hot bath beforehand. Water softens the skin, making it easier to massage, and makes it more relaxing and pleasurable for the receiver.

Use lotion or mineral oil. This makes it easier for the masseuse and more enjoyable for the receiver. It's nice to moisturize your own hands as you go if you're the masseuse. If you're the receiver, wear socks to bed to keep the moisturizers working over night for softer feet (but remember to rub off excess before you step into the shower).

Do not use too much lotion or oil, as this will make the hands too slippery to apply adequate pressure. The result can be almost ticklish, which is not relaxing at all. Start with small amounts, then add more as it is absorbed into the skin. It is important to note that these lubricants will usually not be completely absorbed and will therefore need to be rubbed off afterward with a cloth. We've learned this from personal experience as we've both nearly slipped in the shower the next morning.

Do not stretch the legs out straight. This can cause the knees to lock, inhibiting blood flow and causing tension. Instead, the knees should be slightly bent for maximum comfort. Use a pillow under the knees or place the feet in your lap.

It is important for the masseuse to sit in a comfortable position—you don't want the experience to be miserable for either one of you, as then it will be less likely to reoccur. Sitting on the couch or propped up in bed with pillows, placing the feet in your lap is the optimal position for us. Being comfortable increases the likelihood that the massage will continue for a nice long time. Massaging during a movie will also keep your hands occupied and lessen the desire to snack.

When massaging, use even, steady pressure. Too light and it'll tickle, but too hard and it can be painful. Be particularly careful when rubbing a pregnant woman's feet, as too much pressure can actually induce labor![13] (Research varies on whether this is true, but the spot is believed to be just above the ankle.) With diabetics and those who have high blood pressure, you must be careful not to aggravate their condition with excessive rubbing of the feet.

To produce the best physical benefits, rub all areas of the foot. Massage the top, the sides, and even in between the toes, giving extra attention to each individual toe as you go. Apply a little extra effort to the pressure points, which are the heel and the ball of the foot (where the solar plexus reflex is located). This will result in maximum stress relief and relaxation.

Give equal attention to both feet; the body abhors asymmetry. Rotate and repeat in regular intervals to obtain the perfect balance of pleasure. It is in this way that the receiver will be more likely to reach homeostasis, which is when all the body's systems are working in harmony for the ultimate feel-good experience.

For a truly romantic evening, add soft music and candlelight. Slow, gentle massage will induce relaxation; quicker, more intense movement will induce stimulation. Depends on what you want—sleep or sex! The release of oxytocin paves the way for either.

Massage truly is an unselfish gift of love. For the stressed-out firefighter it is especially so, as many fall victim to sleeping pills, pain pills, and alcohol in their desperate efforts to find pain relief and sleep. A little massage—foot or otherwise—is nonaddictive, has no negative side effects, is free, and is even healthy! And it will have a positive impact on your marriage. What a great alternative.

SEX: THE ULTIMATE HEALING TOUCH

Men and women who have recently had sex respond better to stress.[14] This is believed to be partially due to the pleasure of it, but also due to the *chemistry* of it. A study was done on rats; those who mated regularly had a much higher rate of cell proliferation in the hippocampus, which is the part of the brain linked to memory. They also had more brain cell growth and a rise in the number of connections between brain cells than those who did not mate regularly.[15] This type of brain stimulation is believed to help with positive thoughts and self-control, two necessary elements in handling the intense nature of firefighting.

Men who have sex two times a week cut their risk of heart disease in half (this rate was not as high in women because a woman's arousal is not dependent upon blood flow.) For both men and women, sex can make you live longer (as much as 8 years) and look younger (as much as 10 years). Sex increases blood flow, and releases positive hormones (such as dopamine) that directly counter the negatives of trauma and stress which can age the body and the mind.

When words are hard to find, use the next best thing to communicate your steadfast support of one another—your loving caress. Touch is essential for any marriage, but especially so for the firefighter whose battle with trauma is ongoing. Hold hands, hug and kiss often. Give each other massages and make love regularly and your relationship will not only survive the impacts of trauma, but thrive despite them.

THE HUMAN TOUCH

On his deathbed, world-renowned chess master Bobby Fischer reportedly spoke these final words: "Nothing is as healing as the human touch." That heartfelt sentiment was in response to, of all things, a foot massage from a friend. I could write page after page of examples where touching a fellow human being brought comfort, relief, and hope. And I will get back to the whole foot massage idea a bit later.

If touch is so powerful, then imagine the impact of touch between husband and wife. What energy is present when two people who have committed their lives to each other connect on a physical level? We feel that blessed occurrence does not happen nearly enough. That something so simple and elegant could be relegated to an occasional occurrence is very sad.

Our pages in this book are filled with discussions about things like stress, fear, danger, anxiety, depression, pain, trauma, hurt, separation, and on and on and on. These challenges are heightened in the world of fighting fire and the world of the families who must deal with those challenges. It should provide a degree of comfort that there are some answers that are neither complex nor costly. Some answers are not only proven, but also mutually beneficial in the relationship. Touch is one of those.

TOUCH AND SEX AND...

A few aspects of touching your mate should be obvious.

It doesn't have to take a long time. Many wonderful instances will be brief but incredibly powerful.

It doesn't need to be combined with words. Though words may help at times, touch and silence are powerful medicine.

It does need to be wanted. There is nothing more off-putting than unwanted touching.

Oh, and it doesn't have to be about sex. But it can be, and it might lead to sex more often than you could ever imagine.

One of the best forms of touch remains the good old-fashioned hug. It is comforting to just be wrapped up in a loved one's arms and held tight. There is a solidarity that is felt, a combined strength. On a little wall plaque in a store we visited, I saw this quotation: "Sometimes all you need is a hug from the right person and all your stress will melt away." We all can relate to some form of this thought. And let me remind you, dear spouses: you are the right person. You are the one who can intercede and help alleviate the stress or minimize the fear.

Anne wrote a bit about cuddling so I'll keep this part brief. Whenever we teach on this topic and she mentions cuddling, I only have to wait for a few seconds 'til some firefighter in the audience makes a comment like, "Awwwww, how sweet." After the typical laughter, I'll usually shoot back something like "Laugh it up, but cuddling can often lead to a nice wrestling match." And that is true. Whether that wrestling match is just playful or ends up as more, the connection is fantastic. There's a shared joy in caring about each other and touching each other in ways that no one else can. It harkens back to the days of dating when every touch was electric and special. That stuff is critical to what brought us together and critical to keeping us together, especially considering all the challenges the fire service brings to drive us apart.

AND YES, MASSAGE....

Finally, just a word or two in praise of massage. There are plenty of other types of touch and you can figure out most of those if you pursue this with purpose. I know you're wondering where sex comes into play, but there are whole other chapters on that. Massage is simply one of the finest, most unselfish, sacrificial gifts you can give your spouse. It encompasses so many aspects of what Anne and I teach. There is touch, healing, intimacy, connection, relief, pleasure, and many more truly beneficial aspects that I don't have space to type.

Is it always as pleasant for the one giving the massage? Not really. Most unselfish acts require some give. Is it an inexpensive, effective, and memorable way to help your spouse in all the ways we've talked about? Oh, yeah.

I always smile when Anne brings up the foot rub when we are teaching. Some of the uncomfortable looks in the audience or the outright body language of "That's not happening," are comical. One of my friends who I only see on rare occasions always says to me: "I'm still waitin' on that foot rub." So no, it's not easy or even natural to some folks. But please believe me when I tell you it is effective. I do not believe in most diet/fitness/health/wellness fads that come and go. But foot rubs are a seriously good medicine that feels nice, helps in numerous ways, and gives you a thankful heart toward the one giving the rub. It's always good to do stuff that elevates the feeling of gratitude for a spouse.

The last thing I'd say is to remember that all of this should be reciprocal. It is possible that the firefighter may need this more than the spouse, but every job creates pain, stress, fear, anxiety, and so on. How cool would it be for your spouse to come home from a long day of teaching, for example, to find you offering a foot or neck massage? This is especially true if you're putting some of the other things we suggest in to play. How about this for a brief scenario:

You come home from a long shift at the firehouse, having been up most of the night, and your spouse greets you at the door with a smile. Seeing you are blasted, your spouse says it is up to you whether you talk, take a bath, have a bite to eat, or go to bed. You are hugged as he or she heads out the door to work or school. After you get caught up a bit on rest, eat lunch, and do a few chores, your spouse returns. Because you've been given the blessing of re-entry time, some rest, some benefit of the doubt, you are much better able to give it back. You greet your spouse with a hug and ask what he or she would most like next. How about a foot massage for the aching feet of someone who stood in a bank teller cubicle all day? Or maybe taught kids in school or walked a beat on foot patrol? Do you see the point? I could recreate a thousand scenarios that all come back to doing for your spouse as your spouse does for you. And the best model would be doing for your spouse *before* he or she does for you.

Touch is crucial. It is essential. Its absence is noticed and will increase the horrible possibility of it being found in other places, from other people. You and you alone have the obligation and the honor to aid your mate in the ways described. And we have no doubt you will see benefits that will exceed what you ever thought possible.

NOTES

1. Remarque, *All Quiet*, 123.
2. Bessel A. van der Kolk, James W. Hopper, and Janet E. Osterman, "Exploring the Nature of Traumatic Memory: Combining Clinical Knowledge with Laboratory Methods," *Journal of Aggression, Maltreatment and Trauma* 4, no. 2 (2001): 28, https://doi.org/10.1300/J146v04n02_02.
3. James A. Coan, Lane Beckes, and Joseph P. Allen, "Childhood Maternal Support and Social Capital Moderate the Regulatory Impact of Social Relationships in Adulthood," *International Journal of Psychophysiology* 88, no. 3 (2013), https://doi.org/10.1016/j.ijpsycho.2013.04.006.
4. Karen M. Grewen, Bobbi J. Anderson, Susan S. Girdler, and Kathleen C. Light, "Warm Partner Contact Is Related to Lower Cardiovascular Reactivity," *Behavioral Medicine* 29, no. 3 (2003), https://doi.org/10.1080/08964280309596065.
5. Ibid.
6. Vera Morhenn, Laura E. Beavin, and Paul J. Zak, "Massage Increases Oxytocin and Reduces Adrenocorticotropin Hormone in Humans," *Alternative Therapies* 18, no. 6 (2012), https://pdfs.semanticscholar.org/ec56/c9f2f08e6e39c0d321b551153de1668be455.pdf.
7. T. Field, M. Hernandez-Reif, M. Diego, S. Schanberg, and C. Kuhn, "Cortisol Decreases and Serotonin and Dopamine Increase Following Massage Therapy," *International Journal of Neuroscience* (October 2005), 1397–1413.
8. B. Kaada and O. Torsteinbø, "Increase of Plasma β-Endorphins in Connective Tissue Massage," *General Pharmacology* 20, no. 4 (1989): 487–89.
9. Mark H. Rapaport, Pamela Schettler, and Catherine Bresee, "A Preliminary Study of the Effects of Repeated Massage on Hypothalamic–Pituitary–Adrenal and Immune Function in Healthy Individuals: A Study of Mechanisms of Action and Dosage," *The Journal of Alternative and Complementary Medicine* (August 2012), 18(8): 789–797. https://doi.org/10.1089/acm.2011.0071.
10. Grace Covelli, "Benefits of Foot Reflexology," Livestrong.com, August 14, 2017, https://www.livestrong.com/article/25602-benefits-foot-reflexology/.
11. Ibid.
12. Alex A. Kecskes, "Neurohormonal Effects of Massage Therapy," Pacific College of Oriental Medicine, https://www.pacificcollege.edu/news/blog/2014/11/08/neurohormonal-effects-massage-therapy.
13. Carolyn Williams, "10 Ways to Induce Labor," Livestrong.com, October 6, 2015, https://www.livestrong.com/article/18479-induce-labor/.
14. Andrew Goliszek, "The Stress-Sex Connection: How to Prevent Stress from Ruining Your Sex Life," Psychology Today, December 22, 2014, https://www.psychologytoday.com/blog/how-the-mind-heals-the-body/201412/the-stress-sex-connection.
15. Charles Q. Choi, "Sex Boosts Brain Growth, Study Suggests," Live Science, July 28, 2010, https://www.livescience.com/6776-sex-boosts-brain-growth-study-suggests.html.

PART IV

SEX: A VITAL COMPONENT OF MARRIAGE

Sex and Marriage

When I was asked to write marriage columns for Firelife.com, the first thing my husband Mike said to me was, "I can't *wait* to read the sex column!"

"I am *not* writing about sex!" I declared emphatically, my face flushing at the thought.

"But you have to," Mike argued, "if you want to be at all *real* about relationships." He had a point.

This conversation is so typical of men and women; men just want to skip straight to the sex while women are often embarrassed and try to avoid the topic all together. Can you feel me blushing as I write these words? Mike cheerfully continued to tease me on this topic as he added, "And I'm curious to see what picture you'll use!"

I'm including chapters about sex because it is, as Mike pointed out, a central and authentic aspect of marriage. But please do not cringe in horror, as I am not going to mention any personal details about our sex life. These, I believe, are best kept private and are certainly not to be read about in a book, nor should they be shared at the firehouse. Intimacy is a sacred trust between husband and wife so I'll just keep this clinical with this one personal exception: I do believe that a happy marriage must include a healthy, mutually satisfying sex life. And although we joke about it, sex is a serious marital issue—one that begins with a sacred promise.

When we marry, we make powerful declarations of consent before God. We make vows. We make commitments. We willingly agree to love and to honor during good times and bad, till death do us part. But the vow that makes marriage different from any other relationship is the one we often think least of: to forsake all others and cleave only to each other. To forsake all others; what does that mean exactly? We know it has something to do with monogamy, but it is so much more. To forsake means to leave, to abandon, to renounce, and to turn away from *entirely*. Turn away from what? Sex with other people. This is simple yet complex, because in saying this, you are trusting that one person to meet your sexual needs and desires—for life. And they, in turn, are trusting you to do the same. Other people can fulfill our need for friendship, love, conversation, and entertainment, but only a spouse can fulfill our sexual need. The marriage does not end when we seek others for companionship; it does, however, cease to be a working marriage when we seek others for sex.

This is no small thing. The weight of this responsibility must be taken very seriously. Your spouse's sexual needs are entirely on you. If you don't meet them, who will? Because of the vow, because of the promise, they have no other options. It has been said for centuries: problems in the bedroom will often end the marriage. I recently watched a movie that dramatically portrayed this concept, *Cat on a Hot Tin Roof*. It is based on a Tennessee Williams play and stars Elizabeth Taylor and Paul Newman. Big Mama (Paul Newman's mother in the movie), confronts his wife, Maggie (played by Taylor), about their marital problems. The whole house knows Brick (played by Newman) has been sleeping on the couch. Big Mama, patting a bed, says, "When a marriage is on the rocks, the rocks are there. Right there."

This is so true, for the culminating act of deepest intimacy is indeed expressed in lovemaking. When it's there, the marriage is great. When it's not, the marriage suffers and can eventually end up on the rocks.

One partner for life? Sounds so restrictive, limited—even a bit daunting. In our oversexed culture, monogamy has gotten a bad rap. Why even bother with marriage when it is often so difficult? Promiscuity makes promises of freedom, of more fun, of excitement with no responsibility or hassle. But the truth is, promiscuity fails to deliver on these promises.

Countless studies have found that sex *without* love damages the heart, the mind and even the body. As we mentioned in the previous chapter, having sex two to three times a week *within* a loving monogamous relationship or marriage lowers blood pressure, relieves stress, and even boosts immunities by increasing immunoglobulin A (or IGA) levels by as much as 30%.[1] Ironically, having multiple sexual partners has been shown to raise blood pressure, increase stress levels, and decrease IGA levels rendering the promiscuous even more vulnerable to sickness and disease and depression than people who have no sex at all.[2]

Sex in a loving marriage increases blood circulation, which lowers blood pressure, reducing the risk of heart disease. Firefighters have the highest risk of heart attack among first responders. So if you love your firefighter, love your firefighter. Men who have sex two times a week cut their risk of heart disease in half. In fact, sex has been shown to be so good for the heart that doctors are calling for a screen to include men's sexual activity when assessing their risk of heart disease.[3]

There is an emotional and intellectual safety that monogamy helps ensure, preserving the physical oneness that is at the heart of a beautiful marriage. Marriage is dependent upon your willingness to be faithful to each other and faithful to your understanding of the marital bond. Unfaithfulness to either is a betrayal of your covenant. Constant and continuous obedience to your vow will increase the likelihood that your marriage will be blessed, your home will be a place of peace, and your relationship will be one where you both grow in love. But it begins with understanding and recognizing your spouse's sexual needs and why meeting those needs is so central to a healthy marriage.

It's important to remind our readers that this is our experience from a lifetime of working with couples in and out the fire service. Some of this will certainly fall into the category of opinion, but all is based on experience and supported by research and materials that have greatly benefited our relationship. If your situation does not seem to fit, apply what works, and fill the gaps in ways that are productive. Just make sure it is actually working and not just the latest fad. There is respectful disagreement in many areas of discussion on this topic. We're committed to sharing what we've found to be true and hope it is helpful in providing answers or stimulating a search for those answers.

In his book *His Needs, Her Needs*, Dr. Willard F. Harley Jr. polled 15,000 men and women to determine their top needs within a marriage. The results of these polls, combined with his decades of marital counseling experience, determined a pretty consistent pattern of men's and women's needs, with five standing out for each. Guess what the men picked as their number one need within a marriage? Yes, you guessed it—*sex*. "To the typical man, sex is like air or water. He can't do without it very well," Harley writes.[4]

When a man marries, he promises to be faithful to his wife for the rest of his life. He trusts her to fulfill him sexually. If she is unwilling or unable to meet that vital need and if he's moral, he's literally up a creek without a paddle. Unlike air and water, a man *can* live without sex, but not happily. So fundamental and powerful is his desire for sex that he may be ultimately driven to have an affair. At the very least, it renders him vulnerable to one. Remember, in most cases sex is not just *a* marital need to a man, it is his *number-one* marital need.[5]

What about women—do they need sex like air and water? No, they typically do not. Among the women polled, sex didn't even finish in the top five. Several studies done in this area indicate that a woman's sex drive is lower than a man's.[6] Young or virile married men wish to have sex once or twice a day, while married women typically only want it once or twice a week. Men are aroused quickly and easily, and sometimes even against their will. What stimulates a man? Just about anything, but generally they are stimulated visually.[7] The more fit and healthy they are, the stronger their sex drive, as their testosterone level is higher and their circulation is better. Firefighters tend to be very fit. In the book, "Handbook of Individual Differences in Social Behaviors," sensation seekers gravitate to high-risk jobs and activities such as "firefighting and mountain climbing." Sensation seekers are those who love excitement, challenges, risks, and lots of sex. So there you go; most men have a very high sex drive—male firefighters, the highest of all! (This study also indicated that they prefer rock-n-roll to any other music and movies with lots of sex and violence. Just FYI.)[8]

Returning to Harley's poll, what did he determine to be a woman's top marital need? Affection. According to Harley, women need affection the way men need sex—it is core, central, and vital to their well-being.[9] Affection includes attention, interest, warmth, and tenderness. It includes conversation, but not just words—*kind, affirming words*. And hugs, lots of hugs. Hand-holding, flowers, holding the door—these are all gestures of affection that speak volumes to a woman. My favorite token of affection is a simple little love note left in the morning when Mike leaves for work. So precious are these to me that I save them! The women I know and have spoken with on this topic have told me they love phone calls from the station just to check in and see how our days are going. Show interest in her thoughts, her activities, and seek her opinion. Treat her as you did when you we were courting and pursue her as if she must be won again and again. It is in this atmosphere of affection and emotional connection that a woman desires sex: if a husband showers his wife with affection, she is more likely to become sexually aroused just at the thought of him. For women, sex is more of an emotional connection than a physical one. Women must choose to be aroused; it typically does not happen visually as it does with men (though visuals help).

So deep is a woman's need for affection that if she is not receiving it at home, she may be driven to find it elsewhere. She will be drawn to the friend or coworker who lights up when he sees her, puts his arm around her in a friendly gesture, and hangs on her every word. "Just as a man is more vulnerable to an affair if his sexual needs go unmet, so will a woman be vulnerable if her needs for affection go unmet" writes Harley on affection. If you are withdrawn from your wife emotionally, do not expect her to respond to you sexually. For the firefighter this may be a problem, as the traumatic elements of this job can leave him emotionally withdrawn, causing him to have trouble connecting with his wife. With his heightened sex drive, the cycle of unmet needs can be particularly brutal on the firefighter marriage. As with all we write, if the roles are reversed and the wife has the higher sex drive or is the firefighter in the family, apply the principles to your situation. This is about meeting each other's needs and thriving.

Sometimes these affairs can even begin at the firehouse, where living in close quarters and sharing dramatic experiences can form a bond that may turn to attraction (which Mike will address in his portion of another chapter). A weak home life may enhance this possibility. An understanding of each other's deepest needs and an unselfish, concerted effort to fulfill those needs is the best way to keep your marriage strong, happy, and intact.

SEX: THE NEED

I think I've made it abundantly clear what both men and women need most from marriage. Harley, however, continues on with this point when he adds that sex alone is not adequate for men: "A man cannot achieve sexual fulfillment in his marriage unless his wife is sexually fulfilled as well. While I have maintained that men typically need sex more than women, unless a woman *joins* her husband in the sexual experience, his need for sex remains unmet."[10] In other words, a wife does her husband no favors by just grudgingly acquiescing to his desires; he truly wants her to desire him in return to be satisfied completely, to feel the deepest level of intimacy with his wife. Score another point for monogamy; men typically do not enjoy empty sex any more than women do.

In order for a woman to be fulfilled sexually, she typically must first feel an emotional connection through her husband's outward gestures of affection. If a husband understands his wife's deep need for tenderness and affection and romance and gives these things to her on a regular basis, he will have the keys to her arousal. Harley adds that therein lies the problem for many of the couples he ends up counseling: "The typical wife doesn't understand her husband's deep need for sex any more than the typical husband understands his wife's deep need for affection."[11]

CHALLENGES OF THE FIREFIGHTER MARRIAGE

Fig. 14–1. Keep the fires burning at home, not at work.

Frank & Christine Ricci, New Haven FD, Connecticut

Being married for over 20 years is no small feat. This requires commitment, sacrifice, and suffering. That statement is not a cliché or joke about marriage; it is an accurate representation of what our spouses undergo to remain a family unit with us. We need to demonstrate that we value our relationship at home as much as our relationships at work. This requires patience, persistence, and practice. While I have the greatest wife, I am far from a deserving husband.

Chapter 14 — Sex and Marriage

All the attributes that are celebrated and romanticized in firefighters are sometimes at war with our home life. Firefighters are always helping others but sometimes fail to help themselves and their marriage.

One thing that has become clear to me as the president of Connecticut's busiest fire department is that in the makeup of most firefighters, there are a certain number of self-destructive traits that lead to collateral damage. These traits develop from our social norms and the job's drinking culture. Don't get me wrong, these social norms form bonds and are fun, as well as almost necessary to excel in our professional and social circles. However, these behaviors are destructive to home life. These personality traits allow us to operate 10 feet past common sense by ventilating over a fire while standing on a weaker roof, or making an aggressive push down a hot hallway with blinding smoke. Not to mention moving on without a thought after witnessing a horrific or tragic event. We all see horrible things. We see humanity at its best, this is the part we talk about, yet the lasting impact on firefighters is when we witness humanity at its worst, be it rape, murder, child abuse, motor vehicle accidents, deplorable living conditions, or fire deaths. These are all difficult to reconcile and as a service we tend to compartmentalize and move on. This is how we cope and handle the greatest job in the world.

We must be responsible for our actions and inactions at home and away. We can all do better. By no means am I saying that you have to stay home. I am saying you don't have to stay out all night to be mindful of your friends on the job. Try to bond with those with strong marriages and in turn your marriage will be stronger.

I have often witnessed divorced members at work quickly jumping into another relationship or marriage with new energy and excitement and think, What if that same energy had been put into the first relationship? How often do we see the first wife suffer through the hardships of shift work, standing by as we excel and meeting the adversity of this job with grace? Then when things really build up, the relationship fails. It is what we firefighters do, we move to the next thing. Only with marriage, moving to the next thing often brings financial instability while still having to meet the needs of your children.

What if that same excitement in a new relationship could transform your current relationship instead? Think of all the hassles you'd be spared. Treat your first wife like you would a second wife; it is the key to getting out of a rut. Start by putting down your phone, shutting off the computer and TV, and paying attention. In any relationship, new or old, attention to the little things makes the difference: coffee in the morning, a walk together, flowers for no reason, dinner and a movie. How about a weekend away with no kids? Making the effort is worth the investment. As in work, you get out of it what you put into it.

While we were filming a firefighter mayday DVD for *Fire Engineering* at a funeral home, the director said to me, "When people walk into this room they only wish for one thing: time." The wishes he heard uttered were not for status or possessions, not money or esteem, only time. In a marriage, you must spend time.

You need to do the work. You need to listen and hear what your spouse is saying. One of the best things my wife gave me was her insistence that when our son was young, I would curb the extra shifts and overtime. Time at work, shift work, and that second job just to stay in the middle class all impact the family. To meet this request, we made a financial sacrifice. While I was more than reluctant at the time to do this, I have come to cherish the time she encouraged me to take with my son. And with my travel now, I make a conscientious effort to limit it to three times a year, though my personality would put me on a plane lecturing much more. Marriage is about finding the sweet spot and managing your time. You must work to find the time to connect with your wife. Trust me, the sex will be great when you do!

You cannot control what life will throw at you, however you can control how you react to it. It will knock you down and make you believe it is easier to walk away. Most firefighters find it easy to reconcile closing one chapter and moving to the next. This is due to the nature of our work and is a built-in coping mechanism to compartmentalize our lives and respond to the next alarm. Firefighters know death and its finality. A mantra in our service is "Life's too short to be unhappy." However, it is your responsibility to make it work and find the happiness that is in reach if you just try.

Don't ever demonize or intentionally embarrass your partner. The names we all use in anger only dehumanize our partner in an attempt to justify our own selfish behavior. I have learned that it is adversity and how we react to it that defines us. The last 20 years have been filled with adversity; my wife says I thrive off it and often seek it out. Like Andrew Jackson, I was born in a storm and the calm does not suit me. At home, I have not always handled it the right way. We have experienced job loss, the death of our daughter, prevailing in a landmark US Supreme Court case, two significant life-changing injuries on the job, and my being an ass for extended times due to an inflated sense of self-importance. I have been blessed with a wife who grounds me, tells me when I am wrong, and has chosen to do the work and stay the course for the benefit of our relationship and our kids. We also have the advantage of having great examples. Both of our sets of parents were only married once and are still very much in love.

Remember, it is your job as a spouse to put in the time, listen and hear, handle adversity well, and treat her like your second wife.

THE SEX BRAIN

Men and women have different brains. These differences can be noted very quickly in a marriage; most of them are amusing, some intriguing, and let's be honest, a few are a bit irritating at times. Men and women are unique from one another in ways that can often be incompatible, but can also be incredibly complimentary when viewed as such. But the connection between men and women must begin somewhere for these other differences to be discovered and appreciated, and let's face it, we all know where that initial connection begins—with sexual attraction.[12]

In her book *The Female Brain*, Louann Brizendine, M.D., compares a woman's brain to a man's and points out their marked differences in the emotion centers and the sex drive areas. When it comes to emotion, a woman's brain is like an eight-lane superhighway, while a man's is like a country dirt road. But when it comes to sex, we see the opposite: a woman's brain is typically like a tiny little airstrip, whereas a man's sex brain is like Atlanta International Airport, which boards over 45 million passengers a year!

Brizendine asserts that the brain space devoted to sex drive and aggression is two and a half times bigger in a man's brain than in a woman's. Add high levels of testosterone, which strengthens and enlarges brain connections to the sex center, and a man's sex drive becomes three to four times stronger than a woman's. A woman's sex drive ebbs and flows with her fluctuating hormone cycle, a man's remains on high alert 24/7. A scientific study revealed that it doesn't take much to trigger a sexual response in men. Several couples were shown an image of a man and woman talking. The men's sex area of the brain immediately sparked as if the image were sexual, the women's sex area did not, as they saw the image as two people simply talking.

As was stated before, men generally list sex as their number one need in marriage. For women, sex is typically number six, with talking or nonsexual affection being number one. For men, sex usually means love; for women, talking means love.[13] The higher the stress level in a man's life, the *higher* his desire for sex; for a woman, it's often the opposite. The more *stressed* she is, the less sex she wants.[14]

But do not despair, men, for women *do* want and need sex. What attracts a woman to a man? Some scientists define men as chasers, women as choosers, which has strong evidence in nature: The males of the species put on elaborate shows to entice the female; she then chooses which she will accept. This concept can be seen in bower birds. The female bower bird chooses the male who builds the nest she likes the most. Women often prefer men for the same reasons.

Studies show that worldwide, women choose the same ideal qualities in a mate; 10,000 women from 37 cultures were polled and the results showed that they were more interested in material resources and social status than visual appeal. Universally, women want men who can provide, who will stick around, and who will protect; sexual attraction is secondary. Women are also incredibly drawn to or repelled by odors. Women have a much stronger sense of smell than men do and it's most heightened just before ovulation. Cologne, candles—just some food for thought in setting the stage for romance.

The same worldwide study already mentioned was done on 10,000 men to determine their choices for a mate. It revealed that men choose a woman based on her physical attributes and that typically their choices had little to do with either her social status or her financial one. In other words, men choose a mate based on sexual attraction first and foremost, just another indication of how important sex is to a man and why this should not be ignored or forgotten, as over time, the need may become a problem.

SEX: THE SOLUTION

If things have become a little distant in your marriage, step into the breach, be the hero, and try to fill the gaps. Some effort is better than no effort. Even if your attempts seem awkward at first, keep at it; with time, your actions will become more natural. Making love is a privilege, a special event, and should be viewed as such. It must be a priority, so make every effort to keep your love life alive and well.

Buy her flowers or other thoughtful little tokens, "just because." Your gestures of affection are vital. Play some old rock tunes from your days of necking in the car; the pleasant music can help drown out any pesky, worrisome thoughts or distractions. Lock the door so the kids don't interrupt. Set aside time—make a date! Try going to a hotel. Plan your interludes, look forward to them. With the typical firefighter schedule, daytimes are free, so have a little afternoon delight! Change things up a bit by wearing something new to bed or trying different rooms in the house when the kids are away. To lighten the mood, give each other a short massage; massage raises oxytocin levels, the hormone associated with bonding and affection, and this may lead naturally to making love. Oh, and we women love unsolicited help around the house—men, consider it foreplay.

Marriage is an exclusive privilege to mutually meet that need for physical intimacy, for touch, for pleasure, for a closeness that defies all explanation, till death do us part. This is a lofty goal that sadly often fails. So let's fight fire with fire, as this is how we firefighter couples roll. Firefighting is a proactive profession; let's be just as proactive with marriage and the need for sex, and this will help eliminate any chance of drifting apart as a couple. Make your bedroom your secret and safe love nest, a place where intimacy flourishes, keeping you close and your relationship strong. When you're always looking for ways to surprise and delight your spouse, here's the really cool part—it'll come right back to you. There is no such thing as "free love." Love requires sacrifice. Be unselfish and meet each other's deepest needs, and you will have a passionate love affair that will last a lifetime.

This topic is too important to ignore; continue to search for answers in other resources or check out my numerous columns on Firelife.com for more detailed examples and solutions. If sex is still an issue, I highly suggest intensive personal marriage counseling, as your relationship is worth every effort.

NEEDED AND NECESSARY

The need for sex is real, it is important, and it will have an impact on the happiness of your marriage. If you don't believe that, I wish you well and will not say "I told you so" when you figure out I'm right. But you will figure that out, it just depends on how that enlightenment comes.

Since Anne describes various elements that go into the psychology and physiology of how sex impacts our relationships, I'll start off with some simple observations that have proven very consistent as I've watched firefighters and their spouses through the years.

> **1. The need for sex is going to find fulfillment, whether it happens with your spouse or someone else.** The need for the connection that only sexual intimacy provides is too strong and too important to go unmet. I have sat across from so many firefighters who have found out their spouses are having affairs and were absolutely clueless. So many times it's later revealed sex was lacking in the marriage or that they were having affairs of their own.
>
> One of the realities that firefighter spouses need to acknowledge is the enticing atmosphere that seems to surround the idea of firefighters. That image draws people and, whether deserved or not, firefighters will get attention from the opposite sex due to their job. And they'll get affirmation and respect and the opportunity for physical contact because of that image. I'm not condoning pursuing this stuff, just pointing out what is true. If firefighter spouses allow a distance to grow physically, chances are there will be others ready to fill that gap. And if you

are already fighting with your spouse, the emotional situation may provide the perfect storm of loneliness, need, and opportunity.

Firefighters should recognize that the attention you get is fake. It's not about who you really are, as they don't even know that. It's a symptom of the honor and respect that is accorded you because of the history and image of your profession. It's a mirage that should be treated as such and ignored. What should be first and foremost on your agenda is creating an environment where your spouse wants to have sex with you because they find you appealing. Any honest introspection should bring up numerous ways you can meet the needs of your spouse and create a better situation that inspires closeness. Your spouse is just as susceptible to falling victim to someone else as you are, and since you're gone a lot, the opportunity is there as well.

2. Sex should not be a chore. It is one of the great tragedies of marriage that the busy-ness and pressures of life impose on a vibrant sex life. I plead with you to stop letting that happen. Make taking care of each other's needs a priority and make it a goal to have fun. I believe you will find it absolutely shocking how much affection, good will, reward, and enjoyment comes out of a focused effort to meet each other's sexual and affection needs. It will not fix itself in most instances. This should be a deliberate decision that you make together with complete openness as to what you both need. Remember as you go down this road you are treading on very personal, very sacred ground. A person's sexuality is deeply embedded in their innermost being and should be treated with care. Your goal is a lifetime of good sex. Take the time to care for each other in the right ways.

3. Sex will fix a lot of other stuff. I'll get in trouble for even mentioning this, but a satisfying sex life covers a multitude of other sins, whether issues with the kids or housework or scheduling or name it. If you're satisfied sexually, many of the other irritations of life just don't seem all that important. In part, I believe this is because of the stress relief that occurs when spouses connect on such a deep level. It's also because that connection diminishes the frustrations we feel over the continuous array of minor things that pop up by virtue of living in such close proximity to another human being. Stuff we all do that is minor but annoying. It is hard to be angry or frustrated with someone with whom you have such intimate connection. If you don't believe me, try it out. You can read the rest of the chapter later.

4. Get Real. Don't expect stupid stuff from your spouse. Just because you saw it in a movie or cheap porno flick (more on that later) doesn't mean it should be done. I cannot stress how important it is to come to an agreement with your spouse on comfort levels for sexual activity. If you try to force things on your partner or shame them into something they're not ready for, you will watch intimacy fade. This is not a disclaimer against variety or trying new stuff. As Anne has mentioned, that is really enjoyable and can be very satisfying. But my feeling is that anything that creates fear or distrust is not going to be much fun. If you are asking your spouse to try something new or different, ease into it and see what is behind the apprehension. You may sort it out and you may not. A healthy sex life does not have to include every possibility under the sun, and it is far more important that your partner enjoys what you are doing if you want it to continue. I find it very hard to imagine a situation where a satisfying sexual encounter would not lead to the desire for discovering new things. It's a team sport and you need each other to score.

5. Thee before Me. Make it a priority to put your partner's needs before your own. Similar advice is going to be mentioned in multiple areas of this book, but that's because this is what works. Selfishness is ugly, unsexy, and boring. Those three characteristics are not the typical attributes you think of when dreaming of a sexual partner. Please just make the pleasure about them. Insist upon it. Imagine ways you can delight your spouse and make them feel special, beautiful, wanted. If you are doing that, and it's coming right back at you from the other side, all will be right in the world.

6. Sometimes Yes, Sometimes No. In the real world, life does not always fit your desires. By this I mean your spouse is not always going to be in the same mood as you, especially as it pertains to physical touch. Fatigue, pain, stress, and a host of factors can really distract from intimacy. Give them a break. Don't get angry or pouty or whiny or desperate (I'm pretty sure I've tried them all). Have the grace to just accept the mood is not right and look forward to the next time it is. Consider filling the gap with something that will brighten the day, such as a massage or long walk. These moments of understanding go a long way toward creating an environment of trust.

ENJOY EVERY MOMENT

The final bit of advice is a simple admonition to really enjoy each other. Life moves very fast and is incredibly unpredictable. Take it from a firefighter… we know. You simply cannot see what tomorrow will bring, but you do have right now. And right now should be spent loving and caring for each other. Growing closer in every way possible. Sex facilitates a closeness that vastly exceeds just the physical components. Connect with your spouse in a way that only you can and give each other the joy of devotion and intimacy that is unique to marriage.

NOTES

1. Michael Castleman, "8 Reasons Sex Improves Your Health," AARP, https://www.aarp.org/relationships/love-sex/info-06-2011/sex-improves-men-health.html.
2. "Sex Can Boost the Immune System," Scienceagogo.com, April 14, 1999, http://www.scienceagogo.com/news/19990314232205data_trunc_sys.shtml.
3. Sara G. Miller, "Firefighters Face Highest Heart Attack Risk among Responders," LiveScience, November 12, 2015, https://www.livescience.com/52779-firefighter-risk-heart-disease.html.
4. Harley, *His Needs, Her Needs*, 46.
5. Ibid., 50–51.
6. Richard Sine, "Sex Drive: How Do Men and Women Compare?" WebMD Feature, August 22, 2013, https://www.webmd.com/sex/features/sex-drive-how-do-men-women-compare#1.
7. Anahad O'Connor, "In Sex, Brain Studies Show, 'la Différence' Still Holds," *New York Times*, March 16, 2004, http://www.nytimes.com/2004/03/16/health/in-sex-brain-studies-show-la-difference-still-holds.html.
8. Mark R. Leary and Rick H. Hoyle, editors, *Handbook of Individual Differences in Social Behavior* (New York: Guilford Press, 2009), 460-463.
9. Harley, *His Needs, Her Needs*, 37.
10. Ibid., 53.
11. Ibid., 50.
12. Melinda Wenner, "Sex Is Better for Women in Love," *Scientific American*, February 1, 2008, https://www.scientificamerican.com/article/sex-is-better-for-women-in-love/; Bruce Goldman, "Two Minds: The Cognitive Differences between Men and Women," *Stanford Medicine*, spring 2017, http://stanmed.stanford.edu/2017spring/how-mens-and-womens-brains-are-different.html.
13. Harley, *His Needs, Her Needs*.
14. Andrew Goliszek, "The Stress-Sex Connection: How to Prevent Stress from Ruining Your Sex Life," *How the Mind Heals the Body* (blog), *Psychology Today*, December 22, 2014, https://www.psychologytoday.com/blog/how-the-mind-heals-the-body/201412/the-stress-sex-connection.

Pornography vs. Marriage

Sex is a beautiful, bonding component of marriage. As with all good things in life, there typically will arise counterfeits. Fake substitutes that seek to displace that which is good with that which is not. Hardcore pornography comes in many forms, whether in print, digitally displayed, or live. In our current day it is everywhere and easily obtained. Hardcore online pornography is a growing problem in the firehouse, one for which firefighters have been disciplined, terminated, even jailed. Firefighting is a high-stress profession, and because of this, firefighters tend to seek outlets. This is only natural and often necessary. But sometimes these outlets can form addictions such as alcohol, drug use, and pornography. This chapter is a controversial one as opinions differ widely, but as with anything in our book, it is meant to help, even warn—not to judge or condemn.

> *More and more officers and firefighters are accessing pornographic websites while on duty. I can honestly say that a lot of people I've worked with had issues with alcohol, drug use, and sex addiction. There is no excuse for these behaviors, but the stress level is very high, and a lot of firefighters ease stress by acting out in a variety of ways.... Pornography is the number-one vice that challenges our firefighters to remain morally sound and morally courageous while on duty. It has become so destructive that department management had to have legal departments write special orders, directives, and operational memorandums on the nonaccess or keeping of pornographic material in print or electronically.*
>
> —*Stan Tarnowski, fire chief (ret.)*[1]

And in spite of these preemptive and career threatening measures, there are many firefighters still willing to take the risks to access pornographic material. That is a clear indication of how strong the pull of pornography, in all its varied forms, can become for some.

PORN IS EVERYWHERE

Seventy-two million Americans have seen porn at least once in their lives, while 40 million visit porn web sites *every month*. Many therapists report an alarming rise in treating porn addiction. We are now raising entire generations on porn with the current average age of exposure being *just 11*—imagine that, witnessing an explicit pornographic image before your first kiss?[2] Pornography is readily available to anyone at any time—it's quite tempting, especially for men; it is estimated that up to 85% of viewers are indeed men.

So, what's the big deal; it's just a little harmless fun, isn't it? After all, a bit of porn can liven up the old marital sex life, right? Wrong! Hardcore pornography is anything but harmless, and it's rarely just "occasional" because it is just as addictive as any drug; the chemicals that are released into the brain from drug use are *identical* to those released while viewing porn. Two authorities on the neurochemistry of addiction, Harvey Milkman and Stanley Sunderwirth, have this to say about pornography: "It is the ability of this drug to influence all three pleasure systems in the brain—arousal, satiation, and fantasy—that makes it the 'piece de resistance' among the addictions."[3] Another study asserts the same, that pornography is indeed addictive, saying this, "a sexual compulsion can cause physical, anatomic change in the brain, the hallmark of brain addiction. A preliminary study shows frontal dysfunction specifically in patients unable to control their sexual behavior ... diffusion MRI demonstrated abnormality in the superior frontal region, an area associated with compulsivity."[4] Pornography addiction is a drug so powerful it can destroy a family simply by distorting a man's perception of his wife. Many believe it to be the most harmful addiction when it comes to marriage, and there are three primary reasons for this.

1) Porn creates a negative attitude towards intimate relationships. Countless studies and neurological imaging confirm this. Susan Fiske, professor of psychology at Princeton University, used MRI scans in 2010 to analyze men watching porn. Afterward, brain activity revealed they viewed the women in these images as an *object* instead of a person.[5] (To be more precise, the area of the brain that recognizes images as objects was activated instead of the area that recognizes images as people.) Wives of porn addicts often seek therapy, and they report this as their number one complaint; their husband no longer seemed "present" during sex. Intimacy had vanished. Wives often feel as if they've been "replaced" by not just one but hundreds of younger, more attractive women.[6]

Porn movies are consistent in presenting women as either one of four things; mindless playmates, nymphomaniacs, heartless gold-diggers, or someone who enjoys pain. These dehumanizing views of women are demeaning and present them as subordinates instead of sensitive, intelligent, loving persons with a soul.

Wives report a noted change in their husband's treatment of them as their addictions grow; affection, communication, and nonsexual touch evaporate, and since this is the number one need in marriage for most women, it is no wonder that the relationship suffers.

2) Porn numbs you. You may think that porn arouses you, stimulates you, enlivens you, but in actuality it numbs you like Novocain. The more you see, the more your sexual spirit is killed off, making it increasingly difficult to get stimulated. Normal marital relations become boring. Wives of addicts say that their husband's demands for extreme behaviors increase, sometimes even to the point of roughness. "Men are becoming desensitized to it, and are therefore seeking out ever harsher, more violent and degrading images. Even the porn industry is shocked by how much violence the fans want."[7] The things you see in porn are things you've never seen before and wouldn't see otherwise. It becomes difficult to get these images out of your head; once you're exposed you're never the same. What you believed would enliven the marital relationship has instead weakened it; what you thought would make you feel more alive has instead left you feeling empty and unsatisfied. Many husbands lose interest in their wives, and for 68% of couples in which one or more person was addicted to internet porn, one or both lost interest in sex with each other. And pornography, in some estimates, increases marital infidelity by 300%.[8]

3) Porn trades the truth for a lie. Porn stars are a fantasy. Many are barely over 18, their images enhanced or air brushed. Many have had plastic surgery. As mere images on a screen, it is easy to imagine them as mindless or soulless, making them the perfect sex object. The variety is endless to the viewer; all shapes and sizes are readily available. For the addict, the fantasy becomes more appealing than the reality of a real flesh-and blood person, your soul mate, someone who loves you—your spouse. That is the lie of porn; it promises fulfillment but instead delivers emptiness and endangers true intimacy. Many users begin to isolate themselves, lie, and exhibit secretive behavior. They can become literally married to the web, preferring to gratify themselves sexually alone, withdrawing from their partner altogether. It sometimes has the same impact on a marriage as an actual affair; wives feel lied to, cheated on, and replaced by the "perfect woman," the fantasy—someone they could never be—willing to have sex in ways they will never want.

But here's the truth; I've done a lot of research on this topic, and many of the stories and memoirs I've read have proven to be absolutely heartbreaking. The reality that exists behind the scenes, when the camera is turned off, is anything but ideal. And Mike, as a longtime firefighter, has gone to countless runs where he has witnessed firsthand the harmful physical and emotional toll in the real world, some of which he will relate in his portion of this chapter. It is for these reasons and many more that I believe porn stars are to be pitied, not placed on a pedestal as a sexual ideal.

> *"There are many people in our porn-saturated society who think that porn is harmless entertainment. They often buy into the idea that porn stars truly are the insatiable sex-craving gods and goddesses they are portrayed to be. Regardless of all the overwhelming research and countless personal accounts exposing the dark reality of the porn industry, many still buy into the fantasy that the porn industry works hard to build.*
>
> —Fightthenewdrug.org

Marriage offers so many wonderful options that are healthy and true. Porn promises everything but delivers nothing; it is a lie in every sense of the word.

Pornography vs. marriage—which wins? Which will get all of your passion, erotic energy and desire—a computer image or the flesh and blood spouse who loves you back? Will you trade touch for a lonely fantasy? Will you withhold intimacy and fail to meet that need in your spouse? You are one flesh, based on a promise; remember it is a sacred, joyous privilege to make love in a giving, intimate relationship.

NO BRAINER OF NO BRAINERS

I hope you will again forgive my bluntness. I believe being direct is typically the best way to communicate in matters of great importance, and this chapter really deserves straight on, sledgehammer, aggressive firefighter truth. So here it goes:

Spouses, you are total and complete fools if you don't cherish, pursue and enjoy the amazing gift of sex. There are no-brainers in life and then there's sex… the no-brainer of no-brainers. This beautiful thing gets destroyed when allowed to be corrupted by two of its biggest counterfeits: infidelity and pornography.

SEMPER FIDELIS

So, the positives are the main feature and (I hope) an encouragement to answer the calling. There is another side that is equally important to the mandate (an order to cleave only to each other), and it lies in what you said when you agreed to get married. You used your own words, but they conveyed the unmistakable meaning that you would love your spouse, and only your spouse, till death do you part. This person is the one to whom you committed to be faithful, no matter the storms. (A quick note of clarification before the fireworks start: by "storms" I do not mean abuse. No one ever has to settle for this. Abuse is a crime and should be punished. Never accept it in your relationship and never be the one to dish it out.)

Fidelity in marriage is critical, it is essential, and it is a firm mandate. Even if your marriage is having a rough spot and things like sex are not happening the way you'd like, you committed to your spouse. That agreement includes working with them on issues that have arisen between you, and getting to the place where intimacy thrives once again. There is no acceptable reason, no excuse, and nothing positive that will come from seeking sexual fulfillment outside your marital bond, ever.

I think it's important to lay bare the reality of this most devastating of betrayals. Understand that you are taking the highest, most sacred vow you've ever made and throwing it on the rocks of selfish fulfillment. The person you declared to be the center of your world is cast aside to a place of humiliation, grief, and disrespect. Make no mistake, those around you see this duplicity even if they don't say it to your face. How are we to trust someone who takes the faithful trust of their life's partner and crushes it for sex?

Our culture has moved to a place where infidelity is not as serious a thing as it once was. There are websites that offer opportunities for affairs, and certainly the entertainment world seems to celebrate them. I hope you'll take a bit of time to contemplate the phony reality this represents. Being cheated on is one of the most crushing blows the human spirit can ever endure. It's one of the deepest forms of abandonment and heartbreak that you can inflict upon someone. The consequences to your marriage, to your family, and to your friends are staggering.

And give this a bit of thought regarding your accomplice in the affair: if they know you are married, what does this say about their character? How solid a place are you going to be in if that is who you ultimately end up with? Isn't it reasonable to consider that this behavior will just as easily be repeated with someone else? No matter how tempting and alluring the situation may be, or gratifying in ways that may not be entirely sexual, is it worth the damage that is sure to follow?

SHELTER FROM THE STORM

We acknowledge that some of you will be in tough places in your marriage, places where intimacy seems very far off, and a lot of work is needed to bring it back. It is also a fact that we are all human and make mistakes. Our challenge to you is to put safeguards in place that keep you from making the critical mistakes, the ones that are not easily fixed (if they can be fixed at all). I have a few simple things I do that help to keep the idea of infidelity at bay:

1. **Occasionally, I purposely think about Anne and what it would be like if I cheated on her.** I picture her face hearing the news and the tears and humiliation that would be there. This person, whom I love more than anyone else, devastated by a betrayal that she didn't believe possible. It's a terrible image, a gut-wrenching picture. And it's exactly how it would be if ever I broke faith in my marriage.

2. **My guard is always up around other women.** I have tons of female friends and meet many more teaching and traveling. But I do not put myself in a position of being alone with them. No matter how innocent things are and how committed you are to your marriage, remember you're only human. Fallible and fragile. Things can progress far more quickly than you'd ever imagine and this is especially true if you are having challenges at home. Don't put yourself in any situation that has the chance of getting beyond your control.

3. **Be very careful with alcohol.** Most of you realize that you say and do things when under the influence that you'd normally never say or do. This is certainly true of infidelity. I'd be glad if you eliminated alcohol completely, but I realize that likely will not happen. So please, seriously consider restraint in any situation that might lead to actions you'll greatly regret.

4. **Probably of most importance is to have other people to whom you are accountable.** Set up a group of friends and colleagues who will speak to you openly and plainly about what they see you doing. It is always preferable to walk as a team and have others who've got your back. Especially other firefighters. I've seen affairs begin at the firehouse between male and female coworkers that could and should have been stopped by those who were aware. A family is destroyed when you are complicit. And often, so is the firefighter. You're not doing them any favors by keeping your mouth shut.

A DARK STAIN

Make no mistake about it, infidelity is a crime. It's a repugnant, loathsome, dark, and destructive act that is a choice. If you are cheating, be aware that it will be found out. You will be thought of differently by all you know and you'll spend a lifetime trying to repair the damage to your reputation. Your colleagues may not say it to your face, but they will lose respect for you. And your family will be impacted in a way that is difficult to fully describe. No matter what the circumstances, no matter the temptations and their appeal, don't do this awful thing.

PORNOGRAPHY IS SINISTER STUFF

For men, and certainly some women, pornography has a very strong appeal. Because it is so prevalent, it's important for us to remind you we are not trying to preach at you. We believe you can take a good look at what pornography is truly all about and make some decisions that don't have to be based in spiritual or religious thought. Let me offer a few up and see if any resonate with your thinking. And as I do so, let me answer the question you likely have for me: Does it appeal to you? The answer is of course it does. The female form is intoxicating and the variety available is truly appealing. I will battle the same impulses every one of you does, especially with so much of it easily available. The following is as true for me as it is for you:

Pornography is addictive. You know this and study after study says so. This stuff is fascinating, stimulating, and immediate. You don't have to get anyone's permission or go out of your way to be front and center for the display. And that display typically leads to wanting more and varied and extremely explicit images.

Pornography ruins perspective and truth. Let's face it gang, most of us will never look as good as the men and women in this industry. Youth, genetics, and some tricks in filming enhance appearance to a level that will likely make your spouse pale by comparison. Never mind that this would also be true of you when compared to the actors/participants. How different from the joy that comes in cherishing your spouse and their body. Flaws and all, our bodies tend to be a roadmap of our lives. The scars tell a story. The lines and stretch marks and imperfections relate wonderful things like work we've done, childbirth, pains, and pleasures of our story… our life. That may be a far cry from the "perfection" you see on the screen, but it is way better. And it should be guarded and sacred and special. It's your perspective that's being skewed to think that what you see on the screen is what you should have. And do. And deserve. It is a mirage of the deadliest kind.

Pornography takes advantage of others. This is the saddest thing of all, especially for firefighters who wear a badge that signifies protection. Our calling is to stand in the gap and defend those who are hurting, are weak, who can't defend themselves. Spend a little time reading some of the stories and autobiographies of people who fall into this lifestyle. So much of it is simple desperation, naiveté, or worse. The crimes surrounding pornography and abject abuse are stunning. I responded to an EMS call for a woman complaining of severe stomach pain. As we chatted with her, she kept putting her entire hand into her mouth and down her throat to induce vomiting. Her husband told us she had become a serious alcoholic because of her job as an exotic dancer. She felt so much shame up on the stage that the booze was the only thing that could get her through the night. When we asked, incredulously, why didn't he get her to stop, he said they needed the money. With Christmas coming they wanted to buy expensive gifts for their friends and family. I have never wanted to punch a guy in the face like I did that guy. But that's the type of story that surrounds this world. It is dark and sad and difficult to glimpse up close. And it is we who support and encourage it when we indulge in their product.

Pornography reduces you to rank hypocrisy. Now wait a minute, you're thinking. They are free to do what they wish with their bodies just as you are free to partake. Well, this is not about if you can or if it's legal. That issue is decided by law. This speaks to if you should, based on your own code and what you believe to be true about yourself and the world. And, of course, what you hope to be true about your marriage. Let me relay a story that I have heard in various forms, but all true and told to me by the people involved:

While chatting with a senior officer one day about a training session he had just returned from, he told me what the crews did during their down time. They all went to a local strip club to have a few drinks and watch the show. He grew a bit somber as he told me of the firefighter next to him who was really animated and enjoying one particular dancer. He looked at me and said it really bothered him, not because he had anything against the girls stripping, but because this girl looked exactly like his daughter-in-law. Every time he saw this girl dancing, he saw his daughter-in-law and it made him sick. The lesson here is one that has been told time and time again, but needs to keep being repeated. She is someone's daughter, daughter-in-law, wife, mother, granddaughter or friend. We are horrified to imagine our own loved ones being in that position and would do anything to keep that from happening. Our hypocrisy should convince us that, appealing as this stuff is, it violates our code. It makes us less and has a terrible effect on our ability to have true intimacy with our spouse.

SO, WHO DECIDES WHAT'S GOOD AND WHAT'S BAD?

The simple answer to this question is that you, as a couple, make those decisions. What we have tried to show is there are dangers here that get out of control rapidly and ruin both individuals and relationships. Not everyone is affected in the same way, but you had better have your eyes open if pornography is a game you want to play.

We were very surprised at the tremendous pushback we received for writing about pornography and its dangers. Very, very surprised. There are many who feel it is harmless and can be a healthy part of your relationship. And some psychologists argue there is no such thing as "porn addiction." We obviously disagree and feel an obligation to tell it as we see it and provide warning of the potential impacts.

At a minimum, be sure that whatever you are watching or doing is out in the open and known to your spouse. If it is truly a healthy support to the relationship it should not be a secret. Make sure that interacting with this type of material or actions is not prompting you to more severe and aberrant stuff. And, above all else, your spouse should be able to be honest about their feelings regarding anything you allow to impact your marriage. To be sure, pornography, whether in print, electronic media or in the flesh, will have an impact. We hope you acknowledge the potential harm and choose wisely.

DON'T BETRAY THE FAITH

Both infidelity and pornography involve elements of betrayal to our vows. Instead of seeking sexual fulfillment in our beloved spouse, we look for it in other people or mediums. History is filled with the ravages of these two demons, and the wake they've left is filled with broken lives and shattered homes. Both are tempting and offer pleasures that will surpass what you have with your spouse. Both are a lie. Both are destructive to all parties involved. Both can be avoided.

And they are wrong. It's as simple as that.

NOTES

1. Stan Tarnowsky, "50 Dos and Don'ts, Part 1: Command and Leadership," *FireRescue*, October 1, 2015, http://www.firerescuemagazine.com/articles/print/volume-10/issue-10/command-and-leadership/50-dos-and-don-ts-part-1.html.

2. James Emery White, "Pornified," Crosswalk.com, June 6, 2013, https://www.crosswalk.com/blogs/dr-james-emery-white/pornified.html.

3. Harvey B. Milkman and Stanley G. Sunderwirth, *Craving for Ecstasy: The Consciousness and Chemistry of Escape* (Lexington, MA: Lexington Books, 1987).

4. Donald L. Hilton and Clark Watts, "Pornography Addiction: A Neuroscience Perspective," *Surgical Neurology International*, February 21, 2011, https://www.ncbi.nlm.nih.gov/pmc/articles/PMC3050060/.

5. Doug Eshleman, "Princeton Study: 'Men View Half-Naked Women as Objects,'" *Daily Princetonian*, February 18, 2009.

6. Luke Gilkerson, "The Impact of Pornography on Women," CovenantEyes.com, December 17, 2010, http://www.covenanteyes.com/2010/12/17/the-impact-of-pornography-on-women/.

7. Julie Bindel, "The Truth about the Porn Industry," *Guardian*, July 2, 2010, https://www.theguardian.com/lifeandstyle/2010/jul/02/gail-dines-pornography.

8. "18 Shocking Stats about the Porn Industry and Its Underage Consumers," Fight the New Drug, September 5, 2017, https://fightthenewdrug.org/10-porn-stats-that-will-blow-your-mind/.

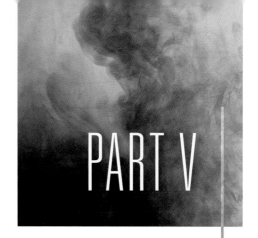

Part V

Children

Kids Are the Icing, Not the Cake

Kids: no matter how much you love them, they will one day leave you. Trust me on this, I know. Mike and I have raised two sons, two fine, wonderful, loving, solid sons. They came unexpectedly into our lives when we were young and turned our world completely upside down. They changed everything for the first 25 years of our marriage, and then they were gone.

Our eldest son, Michael Kenneth, is happily married and has a life of his own (fig. 16–1). He earned a BA in chemistry from Seattle Pacific University (SPU) and is senior medical analyst at Swedish Hospital in downtown Seattle. He has a lovely wife, Samantha (whom he met at SPU), and much to our delight, they have given us our first beautiful granddaughter, Quinn. Michael has his own life and he is quite happy—without us. We're still close and get to spend time and holidays with him, but our son is forever out of our home, building his own family. His childhood bedroom has become a guest room.

Fig. 16–1. Michael's all grown up with his own little family: wife Samantha and daughter Quinn.

Our youngest son, Rick Thomas, also attended SPU and graduated with a double honors degree in English and political science (fig. 16–2). He was accepted into many law schools and chose the one that was about as far away as it gets—George Mason University School of Law (now called the Antonin Scalia Law School) in Arlington (VA), just outside of Washington, DC. He finished his juris doctor and now he is studying for the bar.

Fig. 16–2. Rick's all grown up too with his juris doctor from Antonin Scalia Law School.

The four of us were once magical together, inseparable—close. We did it all: family trips, family holidays, family games, and even nightly sit-down dinners. No foursome could have been closer. Michael was always our "Joy" and Rick was always our "Comfort," for these handles best describe their personalities and the qualities they added to our home. The pain of their leaving and causing our nest to be empty would have been absolutely unbearable if not for one saving grace: Mike and I still have a very happy, satisfying relationship because we always put each other before our children. We never let our marriage suffer because of parenting. As wonderful as they are, our kids have been the *icing* to our lives, not the *cake*. And this was a conscious choice, as it is easier said than done.

WHEN THE SWEETNESS BEGINS

Is there anything more precious, more helpless, and more all-consuming than a newborn baby? I don't believe there is. All my life I wanted to be a mother;

I was the kind of little girl that loved my baby dolls and carried them around with me wherever I went. When the real thing came along, I couldn't have been more overjoyed. But the reality of babies hit Mike and I like a truck: how could something so tiny require so much care, and so much *stuff*? Car seats, high chairs, cribs, diapers, strollers—these items made going anywhere a monumental task. And sleep? You might as well forget about sleep! You think firefighters are sleep deprived? They've got nothing on new parents. The "walking dead" is what I call them (hey, that just might make a catchy title for a T.V. show!) Sometimes sleep disruption can go on for months, causing massive angst. But because of our commitment to putting our marriage before the kids we decided early on that this begins with sleep as this is the first area in which new parents are challenged and where the baby starts to make inroads toward coming between you.

To each their own bed

The battle for your marriage begins in the bedroom, so this is where new parents should decide to do this: keep the bedroom off limits from the babies, and later, from the kids. We decided that once our babies were past the newborn stage (about six weeks or so), they were not to occupy our bed or our bedroom. They must have their own bed in their own room. We believe that the marriage bed is inviolate and must not be breached, even by the children. And my firefighter is such a light sleeper that the babies kept him awake. With a baby monitor you can sleep alone as you can hear them cry without having to have that little body lying right next to you. It will become a habit if you let it—and the longer it goes on, the harder it will become to break. Putting babies to sleep in their own crib is better for you and safer for them because they can roll off the bed or even be suffocated in all the bedding (this may sound a bit overdramatic, but because we are a longtime firefighter couple, we know of instances where these types of tragedies have occurred).

Sex is central to a solid marriage, so you are not doing your children any favors by letting them literally get between you. Easier said than done, I know, as kids have a way of tugging at the old heartstrings and wanting to come sleep with you—but you must be firm and not allow it on any regular basis (the occasional closet-monster scare or illness are a few obvious exceptions to the rule). Babies and children can and will sleep just fine without you.

Bedtime matters

Rule number two on sleeping was this: Bedtime should be strictly observed for the children at every age. Our kids went to bed promptly at 8:00 p.m. all the way up to high school, when it was adjusted to around 10:00 p.m. Teenagers will push the boundaries with video games and movies and such, but parents must continue to be firm. Teenagers face incredible pressures as college looms and after-school activities increase. Be firm when they're young and bedtime will be easier when they're older—when it counts even more. Our boys had such good habits by then that they chose to go to bed at a reasonable time. Because we were strict and consistent about bedtime habits with them from an early age, they adopted these habits willingly as teenagers. And we did not allow our young boys to have a TV in their room—TV overstimulates the brain which inhibits sleep (refer back to "blue light" in chapter 3). They were read to instead until they could read for themselves.

Besides the fact that kids need lots of sleep to be healthy, pleasant, and good students, the other important reason for strict bedtime enforcement was this: Mom and Dad *must* have some time alone for intimacy. This cannot happen if the kids have free reign over your nights, which are already tenuous because of the firefighter schedule. Your private hours as a firefighter couple are more subject to interruption than other couples'; do not let the kids add even further disruption by their presence in your room or by their wakefulness in their own rooms. Sex is absolutely central to maintaining an intimate relationship; setting sleep boundaries with the children will keep them from interfering in this vital aspect of marital health.

The crammed calendar

Keeping your marriage first is basically all about time management; time management that begins first with bedtime, then as the children start to get a life—managing their activity level. It is very easy to overschedule the kids; this is a much bigger problem for parents today than it was for their folks. Many years ago when we were kids growing up in the '60s and '70s, we entertained ourselves. We roamed the neighborhood till dinner. We ran and played, safely, and without adult supervision. We made friends with the kids next door and we went to their houses after school. We walked or rode our bikes everywhere we went, and sometimes we went pretty far. We invented games, and common everyday items became toys. We used our imagination, mostly for good, but sometimes for bad—and even the bad became a learning experience. These things we did without the knowledge or involvement of our parents.

Sadly, today, we parents can no longer let our children run free as the dangers are simply too great. Pedophiles are all over the news; we even get alerts in the mail as they move into our neighborhoods. As a result, we must transport our children everywhere ourselves. We have to check out each playmate and get to know their parents to see if they can be trusted. We then make "play dates." We create safe, structured activities and have to be directly involved if our children are to have any kind of life outside the house. Sports, music, arts, church, friends—these things can dominate the parents' free time.

Add to our already heavy burden of entertaining our children and developing their talents the problem of overcrowded schools—now we must take on lots of time helping with homework if we want our kids to actually become educated. Parenting today can be so daunting and so all-consuming as to leave no time for husbands and wives to be a couple.

Jason & Cassie Hoevelmann, Florissant Valley FD, Missouri

My wife Cassie and I have been blessed with four children. With our busy schedules and so many different personalities and preferences in our family it has been hard to find common interests or time to spend together. This has become even more difficult as our kids have gotten older. But one activity that we have discovered we all really enjoy is *campering*. You may think I said that wrong, but I didn't. That's what we do, we go campering. Our second son, Braden, was a part of the Boy Scouts for a while. I would do some tent camping with him, but from a family perspective, it was something nobody else was all that impressed with, so it didn't catch on. When we bought our first camper and told the kids we were going camping, Braden told Cassie that she couldn't call camping in a pull-behind camper camping, so she came up with "campering" and it has stuck.

We found our interest in campering from attending band camp with our oldest son, Jacob. His school rents a group campground at a state park a few hours from us every summer for their high school marching band camp. More-experienced band parents told us that some families camp there to help out, while most students and parents stay at a motel close by. We had three younger children to take with us that first year, and after hearing that the motel wasn't so great, we decided to borrow a camper from a friend.

Until then, we hadn't realized what we were missing! With no Wi-Fi and no TV, Braden and the girls rode their bikes and played outside, doing kid things while we helped with the band camp. We were hooked.

As it worked out, we eventually purchased an inexpensive, no frills camper (we have since upgraded) and used it to go back to band camp and to soccer tournaments. It allows us time to play board games and puzzles, and while the kids ride their bikes, Cassie and I sit in our lawn chairs and really talk. We discuss our plans for the future. We can be a couple while still being near our children. It allows us to just *be* with few

> distractions. It truly is a chance to unwind and completely relax. And, of course, s'mores are a huge favorite!
>
> My oldest hosted some friends last fall in the camper at our local state park and that was as fun for us to help with as it was for them to spend a weekend in a campground just being friends. We feel like the campering experience brings whoever is there closer and bonds are made and strengthened.
>
> You can bet that we will be taking more frequent excursions in the future. This is our thing, readers, it may not be yours—the point is to find what you love to do that draws you close. Time goes by so fast, but hopefully, this will continue to be a family event when our kids have kids. We'll just need more room for all of us. You'll know us when you see us, we'll have the sticker that says, "We ♥ Campering!"

THE CAKE RECIPE

How do you do it all: be a good parent, but still have time and energy to be a good lover? How do we find the strength and the balance? I must again harken back to the title of this chapter: it begins by remembering that kids are the icing to your life, not the cake. You are a spouse *first*, a mommy or daddy *second*. Parenting is but for a short time, while marriage is meant to be for a lifetime. My dear sister-in-law Melissa once shared with me a wonderful nugget of wisdom on this topic when she said, "Nobody wants to be married to somebody's *mom*." In other words, you should not be so consumed by your children that it defines you entirely, delegating your role as spouse to a lesser position. Children should be viewed as an extension of your love, not the reason for it. Kids will not be around forever, nor do they want to be. Their role in your life is not to fulfill you or to make you complete—that's your soul mate's job.

With this view in mind—that your marriage needs to come first—it becomes a little clearer that the kids must not dominate every free minute of the day. Mike and I accomplished this by limiting our children's activities. They did not play every sport every year. They played one. When they were older and could drive, they then could do more if they wanted to. They picked either an instrument or an art—not both. Don't worry, they'll still get into college even if they don't do 8 million things—ours both did. Family time and couple time must be in the schedule somewhere.

We carefully selected a few friends whom we got to know through volunteering at their schools and whose parents volunteered as well. These families became so trusted as to help us transport our boys around, cutting down on the time needed to be the resident taxi driver.

Limiting TV, computer, and video games is crucial if there is to be any time for homework. In our house, homework came first. Games were confiscated on a regular basis if this rule was broken. I still have a stack of games in my closet that were lost forever! I discovered them the other day and had a good laugh. A more drastic measure, if you are desperate for better time usage, is to get rid of cable. We went without cable TV for eight years and we didn't miss it a bit! We had family video night instead (yes, I know this dates us—we raised them before DVDs existed) and this was a great way to enjoy an evening.

A strict bedtime schedule and limited activities are vital if you're to have any time left over to be a couple. Without it kids can absolutely dominate, and they will if you let them. They have their irresistible charms that can run roughshod over the strongest of wills. Parents don't stand a chance if they're not determined to set boundaries. You, Mom and Dad, are the cake; it is up to you to give the family direction and definition and strength, and this is best accomplished by featuring your relationship. A child's sweetness can turn your "cake" into mush with no substance. Putting the kids second doesn't make you a bad parent; in point of fact, it makes you the best of parents. In seeking to say "yes" to everything your kids want to do, you can become so busy with them that you neglect your marriage. A neglected marriage may end in divorce and be forewarned—this will be truly devastating to your kids.

HOW DIVORCE IMPACTS CHILDREN

Divorce is a watershed moment for a child. It can be, and often is, a domino that knocks kids down and keeps them there. It alters who they might have become in a secure home. I myself am a child of divorce. I experienced firsthand what it is to watch your entire world shatter overnight. To watch something, you dearly loved (your family unit) "die" and not be allowed to even properly mourn its loss. To become someone different than you once were, entirely against your will. Divorce does all of this and more. I still to this day wish with all my heart that my parents had stayed together. It impacted me financially, socially, and even physically; I missed so many days of high school due to stomach pre-ulcers (an unusual condition in a teenager) as a result of stress that I almost didn't graduate (though I had nearly straight A's.)

An enormous amount of research on the effects of divorce on children has been going on for decades now, beginning as early as 1921. Though the stigma of it has lessened of late, the repercussions have surprisingly not changed. Study after study comes to the same sad, unfortunate conclusion: kids suffer mightily, sometimes for life, when moms and dads divorce. A child's most basic needs are for safety and security, which are best emulated by a stable home with established, predictable routines. Family dynamics are the most powerful for good and unfortunately sometimes for bad. Children trust their parents to always be there, but divorce can shatter that trust.[1] Love can end. Nothing is certain anymore. The predictable has been drastically altered, and the two people a child loves most are no longer living in the same home. No matter how civil, divorce causes damage as I can personally attest to; it damaged me.

Though every child is unique, research uncovers some pretty consistent tendencies which are best measured in the form of increased *risks*. Risk does *not* mean guaranteed. It's tough to look at the following statistics, but in doing so it is my hope to remind parents of all the reasons to keep your marriage healthy by keeping it first.

Children of divorce are five times more likely to live in poverty. They have higher drop-out rates and consistently fall behind in math (this is because unlike other subjects, math is cumulative—if you miss one concept you can't move on to more complex ones.) They are more likely to develop drug and alcohol addictions and more likely to become sexually active during adolescence.[2]

Other risks include antisocial behaviors and low self-esteem. Health issues are observed as well; the stressed-out child of divorce gets ill more often and takes longer to recover. These health effects have been measured all the way to the end-stages of life: children of divorce are dying, on average, five years earlier than the children of intact families. This is possibly because of the impacts of poverty and antisocial behaviors that can lead to violence. There is even a possible link to an increased risk for stroke later in life as stress has been shown to alter a child's physiology. (Some of these health effects may be linked to the 39% higher odds of these children becoming smokers as well as the stress.)[3]

Children of divorce are themselves more likely to divorce. They tend to marry younger than average, perhaps out of a need for security.[4] They struggle to have committed, long-term relationships, possibly stemming from an inability to trust. On average, when one spouse is the child of divorce the marriage is two times more likely to fail; when both spouses are children of divorce it is three times more likely to fail. "Compared with children of always-married parents, children of divorced parents have more positive attitudes toward divorce and less favorable attitudes toward marriage. People raised in divorced families are less likely to…believe in marriage as enduring and permanent…insist upon a lifelong marital commitment, or to think positively of themselves as parents."[5]

Children of divorce have a propensity for crime and even suicide. In a recent survey, 2,000 children of divorce were polled. One in ten turned to crime, 8% considered suicide. Some of the reasons for this include an increased exposure to adult aggression, having to comfort a distraught parent, and being forced to lie for a parent.[6]

A friend of mine once told me that if she had known what the effects of her divorce were going to have on her children, she never would have gotten one. She said she'd have moved heaven and earth to make her first marriage work. An intact, happy marriage is the *best* gift you can give your kids. They will be healthier, happier, and more successful in life if they miss out on an activity or two in exchange for sanity between Mom and Dad. If you truly love your kids, love each other *first*. Put a lock on the bedroom door, save some resources for dating, and take time out of each day to reconnect as husband and wife. Be a united front with the children, not one that is easily divided. If you put them *second*, they will truly thank you for this; ours have, many times.

Children are indeed a blessing and a joy that adds immeasurable sweetness. But they are not the structure nor are they the foundation for your life—the marriage is. Your relationship as husband and wife goes on, it remains when the kids have grown. Marriage must come first if you want it to still be sweet when the "icing" is gone.

KIDS ARE COOL, BUT...

Kids are some of the coolest things ever. We had an absolute blast with our boys and they brought stuff to our family that was priceless. We would not have been the same without them and I thank God for the true blessing of being parents. But they are gone now. Not gone in the sense of forever, but grown up, on to their own lives and their own families and their own dreams. And it's just Anne and me again.

That is the way of it, my friends, and that's the way it's supposed to be. Where we see so many problems develop is when the kids come along and become the primary focus and couples lose each other in the process. It's a very easy thing to let happen, as kids can be a dominating force that tends to overwhelm everything else. I hope you'll resist the temptation to let them, as the risk to your marriage is very real.

In some respects, this mirrors many of the warnings we've written about regarding keeping the first family first. In the professional sense, the job cannot be allowed to dominate, or the risk is great that you'll lose your family. Take the same advice and transpose it onto your marriage: keep your first *relationship* first. Otherwise, you may lose it.

What's truly sad about this is that your kids need to see a healthy set of parents modeling what a good marriage should look like. They are going to find something to emulate, as you well know. It can be you, loving and serving your spouse, or it can be one of the horror-show relationships offered up on television. Maybe you'd enjoy watching them find a role model in one of the mindless sitcoms that happen to be the rage?

TAKE CARE AND LOVE THEM, BUT...

Nowhere in here is it suggested that you don't take care of your kids. I think you should really enjoy them. Train them. Nurture them. Give them the tools to take on a very challenging world and contribute to the community. Teach them right from wrong.

But what better lesson can they possibly learn than when you say, "I do," that means your spouse comes first. Always. What better lesson can they learn than the fact that not everything is about them and that their every whim doesn't bring the world running? What better lesson than they are part of a loving, giving, self-sacrificing family that looks out for each other and needs each other to be whole? (See fig. 16–3.)

Fig. 16–3. Riding high on Daddy's fire engine 1994.
Left to right: Michael age 8, Rick age 6.

Chapter 16 — Kids Are the Icing, Not the Cake

TIME KEEPS ON TICKIN', TICKIN', TICKIN'

Probably the biggest challenge you'll face in the modern world of parenting is making tough decisions on how much activity is enough. Here we will likely have a variety of opinions and, certainly, many situations exist that will alter what you do. One simple piece of advice is to remember that the health of the entire family, and you as couple, really matters to everyone. When one aspect of life, whether it be sports, school, hobbies, church, or you name it, is allowed to dominate, something loses. Often a lot of stuff loses, and certainly that includes some of the individuals in the family. We'd counsel a balance that is thoughtful and purposeful. What are your priorities as a couple? What is it that you feel will give you the best chance of raising confident, intelligent, compassionate kids who are using their unique gifts to enjoy life and make the world nicer? That's a great place to begin your plan.

For us, it involved doing some things that were important. Because we had a lot of interests that we felt were valuable, some got left off the plate. As a quick snapshot, here's where our time typically got budgeted:

- *Family time was nonnegotiable.* With my unusual schedule, that sometimes meant skipping school to do stuff together. It meant eating as a family at the table regularly. We played together, prayed together, and had group discussions on a regular basis. Our extended family was also given time, and close friends who were as good as family.

- *Church and spiritual growth were nonnegotiable.* As a couple, we prioritized this for ourselves and raised our kids to be thoughtful and curious about spiritual matters. This included church, but also lessons on values and personal character.

- *School and learning were nonnegotiable.* Even though we joke about being the family whose kids were voted "most likely to miss school," a category created just for them, they only got to do that if they had great grades. Intellectual growth was a featured value in our house. We took our boys to lectures by historians and scientists and visited every museum in our area. We spent the night in the Seattle Aquarium with the fish and seals swimming overhead (my back still hurts from that cement floor.) And boy, did we prioritize reading. So much so that we paid the boys to read books as part of their allowance. The more difficult the book, the better the payoff. There came a point where paying became unnecessary.

- *Play time was nonnegotiable.* We are a fun-loving bunch and we played all the time. Some was with each other, some in organized sports for the kids, and some with their friends. But physical activity and the lessons that can only be learned by participating on a team were a priority. Also, having a coach who is not your parent is a great learning experience (for kids and parents alike).

- *Work was nonnegotiable.* The kids had chores and responsibilities. Some they did by themselves, and some we did as a family. We also did service for others, trying to instill that ethic in our boys' hearts.

Now let me stop there, as the bulleted list could extend for quite a few more pages. Here's the point and I hope you'll get it as you fill in your list: you can't do everything. You can't. You will be running from morning till night, passing each other on the way to the next thing and, likely, not enjoying most of it.

THE PLATE RUNNETH OVER

Please consider leaving a few things off the plate. Pick one or two sports and don't let the year-round domination of athletics take away valuable family time. Serve at your church, but don't let the truly wonderful things a spiritual family brings cost you your first family (we have seen this happen and it is tragic). Enjoy school and learning, but don't let the drive to be on top cost you the precious years of being together as a family. The boundaries you set will help keep time-usage in hand, with something left over for just you two as a couple.

As we began, so we finish.

It is now just Anne and me. If we had not ensured that our marriage came first, the odds of us being the loving, happy couple we still are today would have been greatly reduced. And our kids would still be gone.

NOTES

1. Reinier Bloom, "Effects of Divorce on Children," Children-and-Divorce.com, https://www.children-and-divorce.com/effects-of-divorce-on-children.html#top.
2. Amy Desai, "How Could Divorce Affect My Kids?" Focus on the Family, https://www.focusonthefamily.com/marriage/divorce-and-infidelity/should-i-get-a-divorce/how-could-divorce-affect-my-kids.
3. Lauren Hansen, "9 Negative Effects Divorce Reportedly Has on Children," *The Week*, March 28, 2013, http://theweek.com/articles/466107/9-negative-effects-divorce-reportedly-children.
4. Catherine E. Ross and John Mirowsky, "Parental Divorce, Life-Course Disruption, and Adult Depression," *Journal of Marriage and Family* 61:4, November 1999, 1034–1045.,
5. "Effects of Divorce on Children's Future Relationships," Marripedia.com, http://marripedia.org/effect_of_divorce_on_children_s_future_relationships.
6. Hansen, "9 Negative Effects."

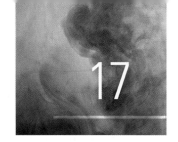

Thoughts on Raising Solid Children

If you're hoping to raise superstars, you'd be better off checking with the parents of Brett Favre or Beyoncé. If you are, however, just hoping to raise loving, emotionally stable, hard-working, respectful, decent kids who graduate from college, then read on. No superstars were produced in our home, but I must say, Mike and I somehow managed to raise two fine young men who've turned out remarkably stable despite growing up amid the chaotic life of a firefighter. Both our boys went to college (fig. 17–1) with academic and leadership scholarships. Neither has ever been in any kind of trouble or has ever rebelled against us. They are moral and self-disciplined and happy. And we've always been close.

Fig. 17–1. Our solid boys, Seattle Pacific University graduation, 2010

So how did we do it? Some of it is just pure dumb luck, some of it's simply who *they* are, but some small portion can be attributed to our parenting. Kids don't come with a set of instructions, but we parents can always look to others for some advice. So here it goes, I'll share a few tips with you younger families that we tried on our kids and that apparently worked.

1. KEEP YOUR KIDS ON A REGULAR SCHEDULE EVEN THOUGH YOUR FIREFIGHTER IS NOT.

This one's tough because firefighters keep chaotic hours that are generally in direct contrast with everyone else's. Mike was often up all night and needed to try to sleep during the day while I struggled to keep little boys quiet. But what kept us sane and connected in the middle of these chaotic, conflicting hours was to be strict in two areas: *bedtime* and *dinnertime*.

During the school years, I had the kids in bed at the same time each night whether Dad was there to tuck them in or not. Mike had one schedule, the kids had another, and I somehow lived in-between the two. And when Daddy was home, we sat down to dinner *together*. I know too many families who *never* eat dinner together and I think this is a tragedy, as sharing a meal is the perfect time to bond. Make it sacred, inviolate, as it's a simple way of keeping in touch so as not to become like ships that pass in the night (or day). We honored mealtime all the way through, even during the teen years. Bedtime and mealtime—two simple cornerstones of stability for the firefighter children.

2. A WORK ETHIC MUST BE TAUGHT BY DOING, NOT BY OBSERVING.

Anyone who knows my boys knows this: they each have an incredible work ethic. I do not believe this happens simply by the parents' example; in fact, the opposite seems to be true. I know a lot of hard-working couples who produced lazy, entitled children who seem to think that money falls off trees because the parents do everything and buy everything for them.

We taught our kids to value work and to do for themselves from a very young age by giving them chore lists. I handwrote lists with actual check boxes, then photocopied them so I'd have a whole stack (this was before the days of home computers with printers.) As soon as my kids were old enough to read, they were given these lists of age-appropriate duties and shown how to complete them. The chores included feeding pets, hauling garbage (with help if needed), even bathing and reading books (my little boys were not willing bathers.) I wanted them to learn to appreciate hygiene and reading and the chore lists helped with this.

When a completed chore list was turned in, they received an allowance, an award for hard work. The money was theirs to do with as they pleased. We typically never just *gave* our kids anything, except on birthdays and holidays. If they wanted something, they had to earn it. They eventually outgrew the chore lists and kept mental note of their tasks, even adding more as they became stronger and more mature. By around age 12 they took on all the yardwork, splitting the tasks between them as they chose; we let them work it out. At age 14, they expanded their earning potential by doing yardwork and dog care for neighbors, extended family, and friends. In high school, they both landed after-school jobs and helped make their own car payments—in our house, if you wanted a car, you paid for it.

The boys also contributed to paying for their education by working during their junior and senior years of college. And they paid their own bills when they lived off-campus. We didn't want them to work their freshman and sophomore years, as we felt that simply adjusting to college life was enough at first. They did, however, always hold down summer jobs.

We told our kids that we were a team, that each of us had our part to do if we were to make it as a family. They had something to contribute and we as their parents needed them to help. And the way we made them *want* to help was by rewarding them with an allowance, with praise, with appreciation, and with respect.

3. PETS HELP KIDS BECOME CARING, TENDER-HEARTED PEOPLE WITH A KEENER RESPECT FOR LIFE.

We all know that firefighters face death regularly—the deaths of others, and even the threat of death for themselves. This reality of firefighting is something we parents strive to shield our children from, and rightfully so. We don't want our little ones to have panic attacks every time their firefighter parent heads to work. Yet at the same time, we want our kids to know a little of death so they can not only be a bit better prepared if the worst should happen, but also to appreciate the value of life.

Enter a pet. A pet is a good safe place to start on this topic. We love our pets, but they're not human, so it's a gradual stepping stone along this path. The first creatures to enter our home were bugs and frogs, a great place to start when learning of life and death as they don't live long, yet they can still be appreciated while they last. Then we got our little boys guinea pigs. They loved these simple, silly creatures and learned how to care for them (with my supervision, of course). We even let our pigs breed, which was another great life lesson. Guinea pigs don't live very long either, but this is the beauty and the tragedy of pets: we love them, we lose them, we bury them, then life goes on. Next, we got a puppy, and our boys grasped an even deeper sense of attachment that was both incredible and heartbreaking. Our family dogs have come and gone (one was even hit by a car), but we believe our affection for these four-legged family members enriched our boys' lives by giving them tender hearts, deeper character, and a better understanding of what their father does for a living—he tries to preserve life.

The firefighter family has unique challenges. These challenges present themselves in a chaotic schedule, the danger of the job, and often tight finances, as firefighters aren't exactly rich. These aspects aren't necessarily negative; in fact, quite the opposite is true—they can be a positive. Time together is more sacred and much more flexible than the typical 9-to-5 routine. Life is precious, and this is made evident by the firefighter's role in society. And because of limited finances, firefighter children can grow up with a strong work ethic if they are shown the way.

Ian & Marita Bland, Queensland, FD, Australia

Ian has been in the fire service for 27 years and has now moved into industrial firefighting. This change has its own challenges with the FIFO (fly in, fly out) shifts. When starting this new adventure, the shift was four weeks on and one week off. After about six months, the new roster changed to 16 days on and 8 days off, which was a far friendlier roster, though still a big challenge for both.

Marita is a registered nurse with 30-plus years of experience with shift work. Throughout this, she managed to raise three children and has been rewarded with three lovely granddaughters for all of those sacrifices.

IAN

Marriage is challenging to be successful at and very easy to be poor at. Add to the mix impacts of two shift workers both working in emergency fields and this brings a few curveballs to the playing field. Our marriage has had some unique experiences: not many could relate to arriving at their wife's place of work doing CPR on a patient and then handing them over. It's not quite the same as handing over a child for a nappy (diaper) change.

Early on in our marriage we decided that our children would never come home to an empty house, but with both parents working shift work, this required some sacrifice, and if the truth be told, this was more on Marita's side than mine. I was part of the USAR team and also a tech rescue officer, so the demands were often a pager away, no different from our volunteers or paid response crews, although when mine went off it was normally for an 8–10 day period. This sudden absence from home meant that Marita covered a lot of my slack.

To help ease these absences I would try to ensure there was nothing that wasn't taken care of, a bit like a prepper's mindset but on a much smaller scale. These were simple things like teaching my boys to hook up the BBQ gas, to mow the lawn, to fix the sewage pump, and to get the generator running. These skills also were just about everyone pitching in to help around the house.

Marita, being of country stock, also had her ways of dealing with these absences: our home always seems to have enough food in it to feed a cast of thousands. During the 2011 Queensland floods, whilst I was living in a wetsuit at the station isolated from my home by floodwater, Marita was caring for the neighborhood with a nightly gathering and BBQ and power-sharing to our neighbors to stop their food from turning in their fridges. I do recall one phone call when the crisis of the evening was "We're getting low on wine and cheese!"

What has worked for us are a couple of simple rules: we try never to go to bed angry with each other, and no TV in the bedroom, so we have to find our own entertainment! When I'm away we always talk, and when I'm home, we share our time with each other.

If I was to offer advice to firefighter husbands, it would be along the lines of remember who your best friend is. She is the one who knows you, warts and all, and still wants you around. Take time off from being the firefighter solving everyone's issues in a 40-minute response time, and focus on being a husband, lover, and friend, father to your children, and a good neighbor. Have fun and don't be too serious at home—we have enough of that at work.

Finally, that issue of manliness. Do I bring work home, talk about what has happened, or do I "protect" my spouse from the job? Early on in my career and marriage I believed I was doing this for all the right (my right) reasons. It wasn't until Marita sat me down and explained that she knew what I did but if I didn't share my day with her, her imagination would start to run. So gentleman and ladies, when you go out to work, make sure you have your "white knight armor" on, but when you get home take it off, it's far more comfortable.

MARITA

Is there really specific marriage advice for firefighter couples? Tell each other you love each other. Often. Kiss hello, kiss goodbye, and kiss goodnight. Every time. You of all couples are all too aware of the fragility of life.

Respect the fact that while you are out there saving the world, she's generally back home saving the family, so when you do get home please remember that she didn't call 911 to get you there, so don't try to rescue her!

It's OK to laugh. Black humor is all you get some days, but laugh all the same.

Housekeeping rules don't change between the travel to and from the firehouse. Your mess, your clean up.

Take an interest in each other's different pursuits. Be proud of what you each bring into the relationship.

Have dates, for even just a few stolen hours while the kids are at school. By not having the consistency of day equals work and night equals home, you miss out on that daily couple time. Use that to your advantage and make those rare times being "normal" count. Get the kids a babysitter if you have to and go out on regular dates. Sometimes I think we were lucky because we had to plan ahead to make all the planets in the universe align to be able to share some private time together. The anticipation adds to the experience. Mind you, in our particular case, our date days more often than not heralded the beginning of some disaster. Mother Nature also used the aligned planets to cast down upon us! We've had major fires, floods, and storms shatter our plans. For some reason I was usually the more disappointed one. Something to do with Ian getting to "go play."

Don't allow yourself to become resentful of the forced time apart that shift work creates. Take advantage of that time to regroup and refocus. Wives are generally the ones to be left on their own, or with the kids, whereas the firefighter can be caught up in a flurry of activity, words, and orders during their time on shift. It can be a clash of needs upon his return home. She needs to talk and catch up and, if she's about to leave for work, do a shift handover, but he needs to sit and sort his thoughts. When the thoughts are sorted, share them with her. Don't unintentionally insult by thinking she's not tough enough to cope with what you did and saw. She married you, how much tougher do you want her to be? Take the time to see where the other is coming from and don't let your selfish need overtake the selflessness of the marriage.

4. CERTAIN HABITS CREATE GOOD STUDENTS.

I believe education begins pretty much from day one. Children who are held close and rocked, talked to, and sung to develop better speaking skills at an earlier age. "Early language and communication skills are crucial for children's success in school and beyond."[1] Mike and I are both avid readers and we wanted our children to be so as well, as we've always believed that readers are leaders and better students. I began reading to my boys as babies, starting with colorful picture books. I taught them their letters, colors, numbers, and shapes. During their primary school years, I read novels to them at bedtime, one chapter a night. They grew to love books and became avid readers on their own. Rick (my law student) even had to be told to "put the book down," otherwise he'd read right through dinnertime. Reading is foundational to being a good student, and it should be introduced by parents. Try reading to them instead of letting them fall asleep to television.

Study habits should be encouraged as well. My boys did their homework before they could play video games or watch TV. They were given as much help as they wanted. I was heavily involved in their studies and monitored their work *very* closely to make sure they were on track and had strong skills, all the way up to middle school, and then I stepped back a bit. I wanted to see how they'd do without me in their face anymore. They began to falter a little, and I let them, as middle school grades don't impact their high school transcripts. I allowed them to experience what it was like to get a lower grade so they'd learn responsibility in time for high school, when it mattered most. By high school, I was not involved *at all* with their studies; my only role was to make sure they did their homework before they did anything else. And to promote their desire to get high grades, we richly rewarded good report cards. You catch more flies with honey than vinegar!

5. IF YOU WANT YOUR KIDS TO HAVE POSITIVE PEERS, GET INVOLVED IN THE CLASSROOM.

Here lies one of the major pros to the firefighter schedule: Dad is available during the day to be a "room mom"! Mike and I both volunteered at our boys' school on a weekly basis; we either fulfilled our commitment together or, if he worked that day, then just I would be there. Our boys loved having their dad in the classroom and so did the kids! Having a man around was a nice change of pace, plus it gave our boys a leg up with the other boys because they too admired Mike (the firefighter).

We wanted to be in the classroom for two reasons: one, to see how our children were fairing, and two, to see what their peers were like. We got to know the other room moms and this helped us pick friends for our kids, as they would be going to these homes to play without us. We needed to trust these parents. My boys' friends and their families were an amazing addition to our lives; we even took trips together! These kids all turned out strong and close and were a positive impact on my children. If you want good influences from a young age, you must monitor what's going on, as children aren't always the best judge of character. And teenagers can be even worse.

During the high school years, we were no longer allowed in the classroom, so we became the "hangout house" after school instead. We didn't want our kids going to homes where there was no parental supervision, so we encouraged our boys to bring their friends here. Another plus to being a firefighter—having a dad around to set the rules. Being the hangout house took extra work on our part—a lot of extra work in fact, as teenage boys eat *a lot* and can be quite destructive (unintentionally usually). But having a houseful of raucous, smelly teenage boys was worth the sacrifice because we knew what they were up to, and besides, they kept us laughing with their crazy antics. They were given lots of freedom and though we were there, we rarely intervened, except at certain times—like the time I saw them loading cement blocks into our fishing boat. When asked what they were doing, they said, "We're going to hold one as we jump in to see how deep down we go." Brilliant ideas like these are why boys, at least, need some adult supervision through the teen years. Though crazy and intrusive and loud and ravenous, being the hangout house was worth it. Trust me, you'll miss those days when they're gone. My house seems so very quiet now.

6. HELP YOUR CHILD HAVE HIGH SELF-ESTEEM AND STAY OUT OF TROUBLE BY FINDING THEIR HIDDEN TALENTS AND INTERESTS.

Everyone has a talent, a gift, something that makes them unique and special. Being a good student is beneficial, but it isn't everything. During the teen years especially, confidence can begin to wane, which is why it's so important to have hobbies well established by that age, so begin finding them early.

Little kids are a blank slate, a bundle of potential, and it's hard to know just what they'll be good at until you have them try. We tried just about everything, from sports, to music, to art, and they were just as excited as we were to do it all. If they showed potential, we pursued it; if they didn't, we let that area slide without telling them they were "no good at it." When one musical instrument proved to be too hard, we tried another until we got a good fit. (*FYI*; studying music is believed to improve math skills.) Michael excelled at piano and guitar and Rick at trumpet and drums. The boys were good at basketball and swimming, but not so good at other sports, so we let them drop.

Talents can include interests and hobbies, such as collecting baseball cards, paintball and airsoft guns, fishing and hunting, and youth group or charity involvements. (I raised boys as you can tell, no girls.) Keep your kids involved in as many areas as possible, especially during the teen years, otherwise they may begin to get bored or impatient. This boredom or dissatisfaction can result in social withdrawal into the "lone-wolf" mentality, or getting into trouble with drugs, even crime. Talent wants to be used; if it's not used for good, it will be used for bad.

7. FAMILY TIME IS DAILY, NOT JUST DURING HOLIDAYS OR VACATIONS.

Because of the erratic firefighter schedule, quality family time can be lost, especially if you're waiting for big chunks of free time to overlap for all. We busy families can push meaningful interaction aside, missing out on golden opportunities each day to connect, hoping instead to get caught up during expensive vacations or holidays. But life is most enjoyed if a little time, just a few focused minutes, is treasured each day. Family life is made up of our simple routines as well as our grander traditions.

For example, make getting out of bed in the morning special with silly wake-up songs—we did. Play I Spy during your daily drives to and from school or to various activities. Share the days' events over dinner—even that can be made into an enjoyable game as each of you takes turns saying, "Today I learned that...." We used to play the "face game" when our kids were little: we'd make a face and the others had to guess the emotion we were trying to convey—happy faces, scary faces, and so on—and we'd have a good laugh at what we came up with.

Every day can be special with your kids in all the little things you do. We still, to this day, love to have heart-to-hearts with our grown sons. We talk of the simple, such as shows we've watched or books we've read, and we talk of grander things, such as our hopes and dreams for the future, or family prayers. Family time comes in many forms, and the most precious memories are often made when least planned. Don't let your family become isolated by television, computers, video games, and cell phones; day-to-day bonding in simple ways is not only free, but absolutely priceless.

8. CHILDHOOD OBESITY IS BEST AVOIDED THROUGH FREEDOM, NOT STRICT RULES.

I know this idea flies in the face of modern conventions, for it seems that people today are more obsessed than ever with what they eat. Every day I hear of new dietary restrictions from the people I meet—"I can't eat this" and "I don't eat that"—yet America's kids just seem to be getting fatter and fatter.

My kids were never overweight; in fact, they were always quite slim. Today they are fit young men and they both work out almost *every day*. One might think I was some sort of food Nazi with my boys, but I was, in fact, quite the opposite. I had two simple food rules in my house: 1.) Mom is not a short-order cook, so you eat what's put in front of you, and 2.) You never have to eat all of what's put in front of you, just stop when you're full. I learned very early on that children have tiny little bellies; they literally *can't* eat very much at one time, so it's cruel to force them to. I let my boys "graze" all day. They were allowed to snack at will, but with my choice of snacks. I did limit sweets and they rarely drank pop (even as adults their preferred beverage is water), but sugar was never forbidden. They had to sit at a table and actually eat dinner with us, but they never had to clean their plates. I genuinely believe that forced feeding and strict dietary rules cause children to become preoccupied with food, which may lead to eating disorders.

Of course, food isn't the only issue with weight: activity should be present as well. With the dangers from predators, today our kids can't run around unsupervised, so you must supervise playtime—make it safe. This requires more effort than ever for us parents, but it's worth it. I did this by sharing the work with other parents: send them to each other's houses, take turns. Kids are more active when they have buddies to play with, and this can happen all year, not just during sport seasons. And because Dad the firefighter had to work out to do his job, the boys learned all about gym equipment from an early age—this possibly helped most of all, because children often follow their parents' lead.

Make family outings include some type of movement—like swimming, biking, or hiking—and this will make all of you strong.

9. IF YOU WANT YOUR CHILD TO FIND A GOOD SPOUSE AND BE A GOOD SPOUSE, THEN WATCH HOW YOU TREAT EACH OTHER.

Researchers at the University of Washington claim they can predict with 87% accuracy which newlyweds will divorce and which will stay together for a lifetime. How do they determine this? Their predictions are based on how a couple talks to each other. Those expressing fondness and love to each other tended to last; those who were rude and critical often divorced.[2] I've known couples like the latter. They're nice to everyone else except their spouse; in fact, their bickering is so prevalent, it makes everyone in the room uncomfortable (and many of these couples did eventually divorce.)

Imagine being the child of these people: they learn that Mom and Dad are hostile to each other, so that's how they should allow themselves to be treated by their own spouse someday. Is this really what you want for your children? Mike and I have always treated each other *better* than others, not worse. We speak well of each other in front of people *and* behind closed doors. Show your kids *by example* what you want for them in their own marriages. Treat your spouse with kindness, friendship, encouragement, and support, and this is what your kids will choose for themselves. This is what our son Michael did. He married his best friend, not his biggest critic, which is exactly what we wanted for him. And he has nothing but praise for his bride, which makes us proud of the kind of husband he is.

10. IF YOUR CHILD DOES NOT RESPECT YOU, THEY WILL NOT TRULY RESPECT THEMSELVES OR ANYONE ELSE.

I believe respect is foundational to good parenting, so much so that if you fail in this area, no matter what else you do for your kids, you've failed as a parent. Parental respect—what does it mean? We know it when we see it, but we especially recognize it when it's absent. You've met these children, and so have I. We avoid them, we don't want to be around them, we may, in fact, secretly wish to be mean to them. I'm referring to the spoiled brat. We watch in horror as these little beasts slap at their parents, defy them, yell and scream at them, and completely disobey with absolute defiance. We think, "Man, that kid's a monster!" But the truth is, the real monster is the parent who has failed to earn this child's respect. They are doing their child a great disservice by making them unfit for society.

Children aren't born respectful—anyone who's had a baby recognizes this truth almost immediately. Babies and toddlers, though adorable, are the most selfish, demanding little creatures on the planet. They genuinely believe that everyone, especially their parents, exists solely to meet their every need—right now! But as early as possible, we parents must educate them otherwise, we must show them, day by day, step by step, that the world does *not* revolve around them. If we do not, they are in for a rude awakening when they go to school. The undisciplined child cannot play well with others or follow the simplest of instructions from a teacher because they have no respect for others or for authority. As an adult, they are headed for disaster in every arena of life from relationships to jobs, and, if not stopped, ultimately with the law.

So how do we do this, turn our selfish little ones into respectful, highly functional children and adults? The first step is to set rules and boundaries. With firmness, not anger, tell your children no. No, we don't hit; no, we don't yell and scream; no, we can't have everything we want the minute we want it. Make the rules clear, define the boundaries, and when they've been crossed, let it be known that there will be consequences.

This leads to the second step: proper discipline. What is proper discipline? Proper discipline is consistent and fair. It is enforced equally with all children. It is done with resolute strength countered by patience and love. And most importantly, proper discipline occurs only when the offender clearly understands the rules, but *knowingly* defies them anyway. For example, we never punished our boys for childish behavior, such as *accidently* breaking or spilling something. We never punished them for forgetting to do something, which is, again, due to immaturity. But we did bring down the law when they deliberately and blatantly defied us, and we did it early and consistently. If you set the boundaries right away and enforce them regularly, they will be honored, because your kids will then know, without a doubt, that you mean what you say.

Teaching your child to respect you and others doesn't end with telling them what's wrong; it also must include showing them what's right. The third step is to earn your child's respect by example. Again, this should begin early. If you don't want your children to yell and scream at you, then don't yell and scream at them. I did the opposite: when my kids were crossing the line, they got "The Look"; no yelling was involved. Since I tried to never yell at my kids, they learned that a raised voice was something to take notice of, as it was not the norm. A loud cry was generally in warning of impending danger, which is, in my opinion, the only proper time to scream at your kids. With "The Look," I could warn my kids that they were crossing the boundaries in public without embarrassing them. Screaming at your children in front of others is disrespectful to them, and it is humiliating. If you do this, you teach them it's OK to shame others. If "The Look" was not enough to deter them in public, then they were quickly removed to the car and taken home. Instead of publicly shaming them, they learned that bad behavior was not tolerated; the outing would simply not proceed. Respect your children's dignity in public, and this will teach them to do the same for you and others.

Show your kids respect in public and, more importantly, at home. Speak to your children with praise and support. Encourage them. Don't call them names, mock them, or criticize them. With kindness and consistency, they will grow to trust you, which is the basis of all respect. If you are considerate of them, they will be considerate of you. If you respect their possessions and give them ownership, they will respect yours. And if you demonstrate character, they will listen to you. And lastly, if you want your children to have respect for others, let them see that *you* do. Firefighters, you have an instant leg up in this area, as few have more honor or care more about people than you. Bring your kids to work, let them see you in action as you serve others, and they will witness firsthand what it means to be a positive, contributing member of society. Firefighters are the antitheses of selfishness; they are truly self-sacrificing. Let your kids see this from an early age and they will admire you beyond words. Your courage will make you their hero for life. This admiration will carry through even into the most disrespectful years of all, the teens. Don't hide what you do out of modesty or a desire to shield them; they need to know about sacrifice to truly appreciate it. You have an ace in the parenting game when your profession is one so worthy of admiration, so use it!

A TRUE HERO: AN EXCELLENT FATHER

The respect our boys feel for their firefighter dad was clearly displayed one Father's Day just a few years ago. In our family, we have established a wonderful tradition: whenever a book is given as a gift, the giver must sign it with a personal note. This gesture turns the ordinary into a memorable keepsake. These are the words our grown son Rick wrote to his Dad:

> Dear Dad; I am the only guy I know who can honestly say that the person I look up to the most, the man I consider to be my hero, is my own father. It's not just because you've spent the last twenty years saving people's lives, fighting fires, and risking your own life to help total strangers; it's not just because so many people now look up to you in awe as a decorated captain and instructor. Mostly, it is because you have sacrificed so much, and given your own time and personal desires so that we, your family, could have a perfect home life. It is because along with your *terrific example*, you have also been all the things a father should be; *a loving presence, a guiding hand, a leader by example, a protector*, and *a spiritual head of the family*. For all this, I thank you. Happy Father's Day! I love you.

Mike hates it when I brag about him, but I'm going to do it anyway. As you can tell by the words written by our son, Mike was an excellent father. I can attest to his loving devotion to his children, as I was there. Of course, moms can do these things too, but this is my chance to speak of my husband and all that he did for our boys. I think dads can learn from his amazing example as both a firefighter and a father. Mike was all of the following and more:

A terrific example

This is step one in being a good father, and firefighters, you are already this simply by virtue of being what you are. As our son clearly stated, children of firefighters admire what their parents do for the community; they are everyday heroes. Firefighters are hard-working, brave, and committed—these are all traits that kids can be proud of. But being a good father does not end there, it takes so much more.

A loving presence

A good father is a tender, gentle, affectionate man that clearly shares his heart and his feelings with his children. He is approachable. This is always important to remember, firefighter dads, as your profession can harden your heart, even to your own children if you let it. In the business of life and death a tender heart is more easily broken, hence the tendency to become "tough." It is therefore even more of a sacrifice for you to remain "soft" for the sake of your kids, but this sacrifice is well worth your efforts, as you will raise emotionally healthy children who flourish in the secure knowledge of your love.

A guiding hand

A good father teaches his children respect through proper discipline. Children will walk all over you if you let them. Undisciplined children tend to be unmanageable, lazy, poor students, and ultimately, dysfunctional and unemployable adults. Mom can do a lot of the discipline, but let's face it, the ultimate authority, the "heavy," the one kids truly tend to fear, is Dad. With but a word, Dad can bring his unruly child back under control. Good dads don't check out when parenting gets tough; instead, they step in, take a hands-on role, and keep the kids from getting out of control, especially through the teen years when limits are stretched to the ultimate. Guidance will keep the kids on course to becoming the best that they can be.

A leader by example

A good father inspires his children to follow his advice if he leads by example. Who listens to those whose lives are a mess? A true role model is someone whose actions are exemplary, even in their private lives. An honest man does not portray himself one way to the world and another way to his family. Children watch your every move; they can spot a phony a mile off. "Do as I say and not as I do" is a recipe for disaster as a father, as the footsteps you try to cover up are the ones your kids will follow.

A protector

A good father makes his children feel safe, not just in the world, but also at home. Dad is strong, Dad is capable, and Dad has my back. He will keep the bad guys away, he will help me fight my battles, and he will be a source of encouragement wherever I go. With a dad like this a child can accomplish anything, as feelings of safety develop self-confidence. Fear, on the other hand, only produces uncertainty, insecurity, a nervous, anxious heart.

A spiritual head of the family

A good father instills morals and values in his children. Proverbs 22:6 says, "Train up a child in the way he should go, and when he is old he will not depart from it." Whatever your religious beliefs may be, it is important to share them with your children, as these views are the ones they will tend to adopt for themselves as adults. If a father is a poor role model, mothers will be forced to look to other men in the family to guide their children, as men have a profound influence on children. A good father does not neglect his children's spiritual needs, as this is the area from which character is developed.

How is a father to do all of these things? The answer to this question is also found in our son's book dedication—it takes *time* and *sacrifice*. A good father gives his time to his children; he takes careful measure of work *and* home life, and does not neglect either. This is where the word sacrifice enters in—the sacrifice of personal pursuits for the sake of family ones. Remember, firefighters, your job is the noblest of callings, one of the finest of professions, and it is one to be truly admired. But even *it* pales in comparison to fatherhood. Your children are your worthiest endeavor.

Mike never sacrificed his children on the altar of his career. He never made me feel like a single parent, not for one minute. It is my opinion that the best husband is also a devoted father, a partner who shares fully in the task. Mike certainly is this. You will find no better example than his—I can honestly say that both his role as a firefighter and as a father make him in every sense of the word, a true hero.

SCARY STUFF, THESE LITTLE ONES

Pretty much the only thing more terrifying than giving advice on marriage is trying to provide wise counsel on raising kids. What a huge, daunting, and immensely important job this thing called parenting is. I think we rather agree with John Wilmot, who wisely said, "Before I got married I had six theories about raising children; now, I have six children and no theories." So please take into consideration the true humility that prefaces these observations respectfully offered on raising solid kids. Here there be dragons and they do bite.

I have one piece of advice that I believe is bomb proof. It is true, will always be true, and should be listened to above all else regarding raising your kids: relax and enjoy them. Enjoy them where they are, for who they are, and please don't rush it. Trust me on this, it will rush itself. Sure, there will be nights that seem as if they'll never end. At times, you will feel cursed by God, inept, helpless, worthless, pathetic, and a total, miserable fraud. Please know it will pass. You will have some wins. You will. You'll have some days with your kids that will challenge for the title of "Best Day of Your Life. It will be a mix and one that you will look back on with great joy if you take the time to be present and involved (fig. 17–2).

Don't strive too hard for the next milestone or development. Enjoy the crawling, as walking will pretty much eliminate that wonderful occurrence. Embrace the amazing sounds that precede actual words, as soon you'll be trying to figure out a way to keep them quiet (and unlearn a few words along the way.) School will come, organized sports will always be there, and don't get me started on things like boyfriends and

Chapter 17 — Thoughts on Raising Solid Children

girlfriends. Please, please, please, just enjoy the ride. As much as you can, soak up the stuff they are doing right in front of you and participate in it as much you can. We can honestly say we worked hard at this and had a good deal of success with being in the moment and not too fixated on what was next. Even so, I'd still pay a lot of money to get down in the dirt with my little guys again and just listen to them jabber on. That cannot be, and it makes me sad. But I'd be devastated if I knew I'd missed it the first time. So, do your best to relax and enjoy your kids because they are very enjoyable.

Fig. 17–2. Raising children was an absolute joy. Our family in 1997 (Michael, age 11, Rick, age 9).

I AM WHAT I AM

As a following thought, while you are enjoying them, let them be who they are. Certainly, provide guidance and shaping, but do your best not to make them fit into your box or do the things you'd wish you'd done or even did. It was a big blow to my ego when neither of my sons really took to sports. I really was excited about the prospect of having athletes and going to big games and rooting them on. But that was not to be.

Now, they did participate some, and it's wise to ensure they get a taste of stuff just to see if they might like it. You'll have to do some gentle prodding. But it quickly becomes clear whether things are in their wheelhouse or not. When it's obvious they are not having fun, are not really learning new skills, or lack the skillset to do well, let them choose whether to continue. If they want to keep hammering away, that's great.

But if they try and then decide to take on something else, go on that ride with support and enthusiasm.

As it turns out, my sons were both good musicians. In an ironic twist of fate, my oldest son developed into an amazing guitar player. My biggest dream when I was a kid was to be in a rock band. That is a problematic dream when you have no musical talent whatsoever. For so long I wished one of the boys would have loved sports as I do. But then it dawned on me, and I told them this, that I'd give up all my awards in sports to be able to play like they do and be on stage rocking out. Let them flow into what lights them up. Just make sure you are providing the environment and encouragement to try new things, be OK with "failing," and assure them they're loved, no matter what.

REALITY CAN BITE

Finally, a reality check to finish off all this stuff that I hope will be helpful and adaptable to your situation and the priorities you set as a family: you can do everything right and still end up with a result that will break your heart. Doing the right stuff increases the odds, big time, that your kids will follow a positive path and become productive adults. But make no mistake about it, your kids ultimately choose their own path and the type of person they are going to be.

I am sad to say that Anne and I have watched some very dedicated and wonderful parents go through absolute heartbreak watching their kids go astray. We know parents who tried every bit as hard as we did to raise solid kids with the following, extremely varied results:

- A couple became professional athletes
- One is a star on Broadway
- More than a few are addicted to drugs
- One threatened to blow up their school
- Many have graduate degrees
- Many barely got a GED
- Some have kids out of wedlock
- Some are married and have kids
- A few are in jail
- Many are in the military
- Some are firefighters

- Some are criminals
- Some enjoy a great relationship with their parents
- Some will not speak to their parents
- Many bring great joy
- Many bring devastating heartbreak
- A few are dead

As with all the advice we give in loving and serving your spouse, we'd encourage you to do the very best you can for your children. And then accept that they have free will and are going to have to make their own choices. For some of you those choices will be difficult to understand and will counter everything you worked so hard to instill. Our job as parents is simply to set the example, light the way, and love them. The rest is up to them, and that's the way it should be. Or as Anne Frank so eloquently observed, "Parents can only give good advice or put them on the right paths, but the final forming of a person's character lies in their own hands."

NOTES

1. "10 Ways to Promote the Language and Communication Skills of Infants and Toddlers," More Than Baby Talk, Frank Porter Graham Child Development Institute, http://mtbt.fpg.unc.edu/more-baby-talk/10-ways-promote-language-and-communication-skills-infants-and-toddlers.

2. Joel Schwarz, "UW Researchers Can Predict Newlywed Divorce, Marital Stability with 87 Percent Accuracy," *UW News*, University of Washington, March 27, 2000, http://www.washington.edu/news/2000/03/27/uw-researchers-can-predict-newlywed-divorce-marital-stability-with-87-percent-accuracy/.

FATHERS AND DAUGHTERS

When my dear father passed away in 2016, his death hit hard and prompted me to reminisce on all that he was to me. I loved my dad; more importantly, I really *liked* him. He was my friend; I enjoyed his company, and now I have one less person in this world who truly loves me. My father wasn't perfect—he certainly didn't do everything right. He often fell short of the ideal as many fathers do. But what he *did* do right is immeasurable. The gifts he gave to me are part of who I am today, and for those things I am eternally grateful. It is these I wish to share with the fathers who are reading this—specifically, fathers of daughters.

This chapter is lovingly dedicated to my dad who passed away before he got to see my book; he was my biggest fan, and I so wish he could have lived to read it for himself. It is, too, written with affection for Mike's wonderful crew, Ladder 5: Rick has triplet daughters. Jason has one daughter. And John has three daughters, as well as a son. Jimmy doesn't have any kids yet, but I'm sure he will someday. I wanted to include something just for them. And most especially, this chapter is for my son Michael, who is himself raising a little girl of his own. I hope you readers will indulge me on this as I know it is a bit exclusive. I feel I covered being the mother of sons, and Mike's being the father of sons, but the father/daughter relationship has been left out because we didn't have any daughters. So, to write this, I will have to borrow from my own childhood.

It's easy to underestimate the role a father plays in his little girl's life. Fatherhood, after all, tends to revolve around sons; at least that's what my father once told me.

Fatherhood with boys is a bit easier and more obvious, he said. Boys need a positive role model and a tough disciplinarian who isn't as easily bowled over as Mom. The rough-housing, vigorous play, and sportsmanship that a man can provide to his sons is part of their development. Only a man can truly impart manhood. He used to ask himself, "But what does a father do with a daughter? After all, I'm not trying to turn my little girl into a man." One of his nicknames for me was "the Wee Little Stranger," as he felt intimidated at the thought of ever fully understanding or being able to relate to his little girl. He told me that his own sister, my Auntie Jan, used to complain how their dad never spent any time with her. She was jealous that he only spent time with his sons, hunting and fishing and drinking and the like. My dad took this to heart and when I was born, he decided he would make more of an effort with me than his father ever did with his sister. And I'm so glad for it, as fathers have a very meaningful influence on their daughters whether they realize it or not.

So what then does his role mean to her? A father's relationship with his daughter is powerful as it impacts her in the following ways:

Mental and physical health: Attentive dads give their little girls *self-confidence*. Girls tend to have to try a little harder to get Dad's attention than Mom's; so, if she has it, she feels empowered. And this empowerment makes her less fearful of men, of anyone really.

Countless studies have shown that girls with loving fathers are much less likely to become clinically depressed, develop psychological problems, or become addicted to drugs or alcohol. And girls with no father are far more likely to exhibit aggressive, even criminal behaviors. (This is true of boys as well.)[1]

Body image is also impacted more by a father than by a mother. Girls with involved, positive fathers are less likely to develop eating disorders. And healing occurs more quickly when Dad becomes involved; for example, anorexic girls recover two times more quickly if spending time alone with Dad is added to the therapy.[2]

No one quite knows why this is so, but girls get a positive or negative body image from their dad. If a father only points out his daughter's physical attributes, such as her beauty or the lack thereof, then she may become preoccupied with just that and suffer immensely if she's not "perfect." But if a father points out his daughter's intrinsic values, such as character, intelligence, courage and integrity, she becomes less focused on her physical attributes and more confident in her personality and worth.

Education and Achievement: A US Department of Education study found that highly involved fathers' children were 43% more likely to earn mostly A's, and were 33% less likely to repeat a grade. This is true for both boys and girls. And another such study found that girls were more likely to stay in school and have higher quantitive and verbal skills, and higher overall intellectual functioning if they had an attentive father in their life.[3]

Fathers influence a daughter to become achievement-oriented, competent and successful in many arenas including sports, music, the arts, and academics. A father's words of, "You can do this" have tremendous impact. For this reason, daughters of loving fathers are far more likely to receive college scholarships. And thus, successfully educated, they are more likely to become successful professionally as well.[4]

Romantic Relationships: Pediatrician Meg Meeker writes in her book, *Strong Fathers, Strong Daughters*, "Fathers are a template for all male figures—teachers, boyfriends, husbands, uncles, and even God Himself—in a daughter's life."

Dr. Linda Nielsen, Ph.D., Professor of Education and Adolescent Psychology at Wake Forest University, adds to this concept when she writes, "Dads, more than moms have the greater impact on the daughter's ability to trust, enjoy, and relate well with the males in her life."

Daughters of loving, involved fathers tend to wait longer to become sexually active and are far less likely to become pregnant during their teen years. They are the most likely to demand to be treated with respect, and thus less likely to be talked into having sex with someone they don't love. Countless studies show that for girls, *less Dad equals more sex*. Deprived of a loving male presence in her life, a girl may be more likely to seek substitute male attention, even if it's negative. The closer she is to Dad, the easier it is for her to refuse unwanted sex.

And most importantly, girls with loving fathers tend to marry better and have longer, happier, satisfying marriages. Accustomed to emotional intimacy with her father, she will settle for no less than that in her mate. Girls are drawn to the familiar; they will typically marry a man who is at least somewhat like their father, whether good or bad.[5]

WAYS FOR FATHERS TO BE CLOSE TO THEIR DAUGHTERS

The biggest mistake fathers can make is not understanding their significance. A father is much more than just a protector or a provider, as was just clearly established in the previous text. Fathers may find it easier to relate to sons, but they must make a concerted effort to find ways to stay close to their daughters, as so much of their healthy development is at stake. Start out attentive when they're little, and stay close even through the teen years. This is the age when it can become awkward for a father and he may start to withdraw a bit. But it is then that she needs him most of all, to guide her through her vulnerable years till she's old enough to truly make wise decisions.

Rick & Karen Owen, Country Fire Authority, Melbourne, Australia

I am blessed to have two daughters, but what makes it better is that they are twins. They are not identical, and each has her own personality and set of traits.

Working shift work has been great but also a drawback, as some shifts occur when important dates are on and we all know how hard it is to get time off.

To overcome this, I now work days and have more flexibility with my time for the girls. After 23 years of shift work, it was time to ensure that the family came first.

To further this, I have tried to find something to do with each of the girls. It has taken some time to achieve this with one of my daughters. She is much like me, very shy at first but once you get to know her, she's a ball.

My eldest twin could spend hours on the screen if we let her, so to engage with her, we play Pokémon Go together. Yes, I know, a grown man playing Pokémon, but this gives me one-on-one time with her. I had to learn a new skill and don't see it as a chore. We go for walks with the dogs, chase and collect Pokémon, and generally get to talk. We chat about a lot of things, sports, school, music, family but hardly ever about my work. If we do, I only share the good experiences; I don't want to burden her developing mind with the sights we see.

The younger of my two is very outgoing. She just wants to know everything. I could be working in the garage on a project with power tools, and she'll want to learn and have a go. Although my garage is my "man cave," my girls are welcome at any time. Now she can use the cordless drills and saws, and understands the safety aspects of each tool. Together we have made many school projects or things for around the house.

What she and I like to do the most together is go for a bike ride. She just loves being on the bike out in front with dad trailing behind. We ride along the local river and bike tracks for miles, and there is never one complaint.

It is a very basic concept, but I have to make time for the girls, be understanding, and most of all, be loving. I sometimes must go outside my comfort zone or do things which some adults may consider embarrassing, but to me it is engaging with my daughters and I get no better feeling than having them smile back.

I spend a fair bit of time traveling for work, local, interstate and sometimes overseas. The most important thing is to phone them. I always try to talk each day, even if it means getting up at 3 a.m. on the other side of the world to call them at home. For overseas adventures, I always make sure that I bring something back for them which can only be obtained from the place I am visiting; hopefully they see it as a reminder of the love, even though I could be 10,000 miles away.

I am very protective of my girls as most fathers are, but as firefighters we witness a lot of other people's traumas. Although I don't ever talk about this with them, the tragedies I have seen only reinforce my love and protectiveness as I don't want them to ever experience the lessons we see from others.

Finally, my daughters drive me to be a better person both in general life and at work. Everyday their presence ensures that I get out of bed and do the best I can at whatever I am doing. At work, we only get one shot at anything and there isn't that chance for a do over, so this attitude also helps in my home life. I only get one chance with my twin girls.

The father-daughter relationship is so precious, so special, so unique. The best fathers find that magical balance between rules and freedoms, closeness and authority. This is hard to do with sons, and it can be even harder to do with daughters. A softer touch is needed. A stretch, an effort, an imagination to relate to that "wee little stranger" who is so very different from himself. But that effort must be made as a father impacts his daughter in the three crucial ways I mentioned: her mental and physical health, her education and achievements, and her romantic relationships—these are pretty weighty things to lay at a father's feet. How involved and close he manages to be throughout her formative years is what makes the difference. But how is he to stay close to his daughter when she is, in fact, a girl, and he's a man? And what if Mom and Dad divorce and Dad doesn't even live with her anymore? Being close is much easier said than done.

My father failed at some things, but he did manage to keep me close in all the ways that matter. Here are four things he did for me that I believe any father can do to help his daughter become the better version of herself. Don't worry if you've never done them before; it's never too late to start.

Activities: My dad took me fishing. I loved doing this with him. It was just the two of us in a boat on a peaceful lake or a pretty river or the majestic Puget Sound. It was an adventure; I was vulnerable, he was strong and capable. He made me feel safe. He'd bring goodies, bait my hook, and we would talk or just sit quietly together enjoying the scenery. I didn't really care if I caught a fish; I just liked being with him. We both derived incredible joy from being on the water; this we had in common.

Small trips together, drives to Idaho (his home town)—we did these things too. And we both shared a love of books and classical music. We shared them, exchanged them—all our lives. He'd read a book then send it my way, and vice-versa. Then we'd talk about it.

Find a common interest with your daughter, such as sports, the outdoors—movies. When you share activities with her, you teach her what it is to be friends with a loving man. And she will expect no less from her husband.

Mini Dates and thoughtful gifts: Yes, I went on "mini dates" with my dad, though when I was a kid in the '70s and '80s, such things were not even heard of. My dad took me to lunch on a regular basis from the time I was old enough to sit quietly in a restaurant with him till the day he was too infirm to drive anymore. A meal in a restaurant with my dad, just like "a date." We'd eat and chat and he'd tell me his many stories and we'd laugh a lot. I loved hearing of his adventures, particularly his spy stories. I learned to be treated like a "lady" by a man who held the door for me, shared a nice outing or meal, and attentively listened to all I had to say. He made me feel as if every word I uttered was worth listening too.

My dad bought me little trinkets or brought me flowers from wherever he went, whether it was a hunting trip or just a trip to the store; he always surprised me. He made me feel special, thought of, treasured. This is what it feels to be "courted," and this is what my husband does for me now. I am worthy of perpetual "courtship" and my father taught me this.

Significant Events: Someone once said, "90% of being a good dad is just showing up." Sad, but true, as so many dads just bail—especially after a divorce. They typically are the ones that move out, and the kids remain with mom. It's easy to lose touch with your kids when you don't physically live with them anymore.

My parents divorced when I was 13 and my dad did indeed move out. But he chose to live close by. He didn't bail; in fact, he doubled his efforts. Our lunches and outings increased. He was there during this major turning point and it helped keep me somewhat sane during a tough time; he simply "showed up."

And dads, be there on your daughter's big days; achievements, graduations, and most importantly—her wedding day. I needed my dad to walk me down that aisle (fig. 18–1); I was young and emotional, and when I entered the church sanctuary and saw all those faces staring at me, the thought of walking past them was simply overwhelming. But on his arm, I faced the crowd and he kept me from shaking—then I saw my husband's smiling face—and all the fear and anxiety and doubt vanished. My father was no longer needed, and he quietly stepped back as I let go of his arm and reached for Mike's hand. My dad "showed up" and his presence gave me strength. Sometimes, that's all your daughter needs.

Chapter 18 Fathers and Daughters

Fig. 18–1. My dad walks me down the aisle in 1985.

Words and Notes: Words are very powerful to girls. Words often elude a man. My dad was great at telling me of his interests and sharing with me his stories. This was a wonderful place to start, as I learned to easily converse with a man.

But my dad struggled with saying he loved me. He was uncomfortable talking about his feelings; this did not come naturally to him. Stoic, sarcastic, elusive; this was his nature. He had a sharp, sometimes biting wit, and this made it hard for him to always speak tenderly. But he found a way to communicate his affection in just as powerful a way—with notes and poems. So impactful were these to me that I have saved them all in a scrapbook I keep.

When I was a child, he slipped loving notes into my lunch box. He gave me cards for my birthdays and other holidays with typed stories or poems, as I could not read his handwriting. (Typed, mind you, on a *typewriter*; we didn't have computers and printers back then.) He took the time to communicate with his little girl that so needed to hear of his love.

On my first day of high school he left me a package outside the front door (as by then he no longer lived with us.) Inside I found an opal necklace and this wonderful poem that he'd written, just for me:

Anne–

An opal is just a stone

Of no great consequence–

Until cut and polished.

Then an iridescent fire

Of many hues and lights

Gleams through the

Haze of alabaster.

Akin perhaps to the often

Bland and placid

Exterior of youth

Until cut by experience

And polished with character.

Then the hues of charm and wit,

The fire of warmth and depth

Also shine through.

Then

Both the stone

And the once baby girl

Become

Truly precious.

(just cause I love you.)

Chapter 18 — Fathers and Daughters

Fathers, tell your little girls that you love them. That they're precious, special, wonderful. If you can't say it in words, then write it down. Find a way; she needs to hear it.

I thrive on kind words; I expect and deserve to hear them from my husband, and I do. And this, too, I learned from my father (fig. 18–2).

Fig. 18–2. Me and my dad in 1967

IT TAKES A MAN

I always joke with my buddies who have girls that it takes a real man to raise a daughter. And that's why God gave me two sons, as He knew I couldn't handle a girl.

Implicit within that joking is a real belief that doing this "daddy thing" for your precious girls is a solemn and sacred and wonderful task. It is different in so many ways from raising sons. Men mostly get what their boys are going through, but have some real gaps in experience when it comes to understanding their daughters.

As I read Anne's descriptions of the things her father did for her, it was humbling and very eye-opening. Some of it I would have never known and can only say a prayer of thanks that those things were done. They proved very meaningful to my sensitive and loving wife and helped shape her into the wonderful woman she is today.

THE EXAMPLE WE HOPE TO SEE

While I do not have a lot to add to this section, the one observation I feel is imperative is for daughters to see their mother being treated well. There is simply nothing that replaces the authentic example of spouses treating each other appropriately. Your kids, and especially daughters watching their fathers, learn how women are supposed to be treated by men. It is devastating to watch negative patterns replicated in so many bad relationships with things such as emotional, physical, or alcohol abuse. Our goal as fathers should be to make the things they see and remember exemplary. Instead of harshness, they see kindness and tenderness. In place of hitting, they see hugging and helping. The escape that things such as alcohol and drugs try to create is instead demonstrated by living life together and enjoying each other's company.

PERFECTION NOT NEEDED

I had quite a few issues with some of the ways Anne's father chose to treat his family. There were issues with detachment, tone and selfishness, that made it difficult for me to see the good at times. But as you read in her words above, there was a lot of good and that good had tremendous impact on who she would become. Myself and our boys are the direct beneficiary of some of the things Ken did show his daughter; love, and a belief that she was special and deserving of respect. I hope this encourages all of you "mere mortal" fathers who know you are not superman—you don't have to do every little thing right. But you must give it full effort and put your heart into the job.

Your little girl(s) are far more important than your stocks and bonds, your sports teams or where the next hunting trip will take you. They are given to you for a short time and plenty of other voices are going to be knocking at the door of their hearts and minds to shape them in perhaps negative ways. Instead of fretting about the day some slovenly kid is going to come to your porch asking her for a date, invest in her character and self-worth so she'll be ready, no matter who comes knocking. Give her a true example of how a lady should be treated by loving your wife in a sacrificial and caring way. Challenge her to walk on her own, knowing that you are there and rooting for her to love life and make an impact with her talents. And make sure she hears from you, in words or notes or gifts, that you love her.

Anne's father did this for her in many, very meaningful ways. And I am so grateful and blessed that he did.

NOTES

1. "Statistics of a Fatherless America," dads4kids.com, https://photius.com/feminocracy/facts_on_fatherless_kids.html.

2. Melissa H. Smith, "Exploring the Daughter and Father Connection in the Treatment of Eating Disorders," Center for Change, https://centerforchange.com/exploring-daughter-father-connection-treatment-eating-disorders/.

3. Meg Meeker, *Strong Fathers, Strong Daughters: 10 Secrets Every Father Should Know* (New York: Ballantine Books, 2007), 24.

4. Linda Nielson, "How Dads Affect Their Daughters into Adulthood," Institute for Family Studies, June 3, 2014, https://ifstudies.org/blog/how-dads-affect-their-daughters-into-adulthood.

5. Ibid.

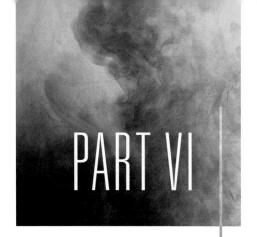

PART VI

MONEY

For Richer or Poorer

Of all the things that could cause the demise of your marriage, did you ever imagine that it might be money? We tend to think that only the obvious land mines, like infidelity, are the ones to worry about when getting married. When you were a young, passionate, star-crossed lover, it was easy to vow "till death do us part." It was romantic to pledge to forsake all others and noble to promise to be there when sickness comes and health fades. Newlyweds, however, tend to just gloss over the "for richer or poorer" part, as these words seem almost mercenary, even ignoble on a wedding day! Love is, after all, a heart issue, not a superficial, materialistic one. Money can't buy you love! That's why it's surprising, ironic, and even a bit shocking to find out later in life that money issues are often cited as one of the top three reasons for divorce in this country. These three revolve around "gaps"; sexual gaps that lead to lack of intimacy or even infidelity; money gaps or differences in spending, saving, and earning habits; and communication gaps of not being able to openly discuss these or any other issues without fighting.[1] A Kansas State University study suggested that arguments over money may be the top indicator of a divorce. Larry Burkett, a leading financial counselor has this to say about discussing finances in a relationship, "Money is either the best or the worst area of communication in (a) marriage."[2] It is either teamwork and shared goals, or a fight waiting to happen. When two people become one flesh, all that they are, including their monetary habits, are now shared. Spenders can marry savers, interests and values collide, and these collisions may be quite volatile. It is true, money can't buy you love, but it sure can cause you a lot of grief!

OUR MARITAL MONEY PHILOSOPHIES

In this chapter, I'm not going to tell you how you should or shouldn't spend your money. I'm not going to give you tips on wise investments or ways to save; consult experts on this topic, as they will be far more qualified than I to guide you. Mike and I are not rich, so you probably wouldn't want my advice in that arena anyway. I can, however, share with you the attitudes or philosophies we have developed over the years that have made money a nonissue in our marriage. Mike and I rarely, if ever, argue about finances. This is due to money philosophies we have found through experience and research to be the healthiest ones to have within a marriage, and conversely, which destructive ones to avoid.

It's not 'your money', it's not 'my money'—it's 'our money'

Mike and I believe that this is the number-one attitude married couples should have toward money. We have always had one joint bank account and regardless of who makes what, it all goes into the same pot. If couples remain separate in this area, it is easier to view it as "mine" and "yours." Viewing it as "ours" is tougher to do when it remains divided after the wedding. When it comes to paying household expenses that are shared, resentments can arise over who should pay which bills if the money is considered separate. You are one household, one family, one couple, and your expenses should also be viewed as such. "We can't keep each area of our marriage neatly separated—money touches everything, so if a couple is fighting about money, that tension can also affect areas like trust, parenting—even intimacy. Not only is it better to be open with one account for reasons of trust, it's easier—ever try splitting bills with a roommate?"[3] The "yours" and "mine" attitude can more easily lead to secret spending, mismanagement, and eventually, mistrust.

It is, of course, necessary to assign the managing of the one account to one person; this doesn't negate the "our money" philosophy. This task should be given to the spouse best suited to handle it. In our household, that person is me. Mike is the primary breadwinner, but he's *terrible* with details. I handle the money, but I do not control it. Mike makes most of the money, but he has never once, in three decades of marriage, referred to it as his. So disinterested is he in the details of money that he doesn't even know how much his paychecks are. He turns it all over to me and trusts me completely with the results. It is ours, plain and simple.

All money problems should be faced together

When financial problems arise, they should always be viewed as "our problem," not "your problem." Casting stones, blaming, and belligerence only exacerbate the situation. All money problems should be faced together, and the sooner the better, as they can quickly escalate. Because I handle the money in our family, I have to tell Mike when there is a problem. Keeping it to myself isn't going to solve anything. He doesn't always want to hear it, but he has to. I cannot always handle the issues alone, nor does Mike expect me to. Our finances affect us both, so we both have to be involved. When difficulties arise, we work them out together. We brainstorm, strategize, and find ways to solve the problem by communicating, not fighting. We figure out where we're overspending and decide how to rein it in, to stop the hemorrhaging.

We've always tried to view our financial struggles as part of the great adventure of life, because one way or another, they will always come. They're not the end of the world. Try to keep it light, and have hope. As G. K. Chesterton puts it, "An adventure is only an inconvenience rightly considered. An inconvenience is only an adventure wrongly considered."[4] Tackling the money mountains of life is doable! You can achieve great things together!

Bigger purchases must always be agreed upon

The larger expenditures, such as houses, cars, and vacations, should always be a joint venture. They need to be 100% agreed upon and well-planned. If they are not, one spouse can quickly become resentful of the purchase. Where's the fun in that? Your bigger investments are supposed to be exciting, to be a source of joy and accomplishment that you achieve as a team, not as an individual. We only get a few of the "biggies" in life, so make them count. If you do it together, you will truly cherish it forever.

Thou shalt not commit financial infidelity

Financial infidelity is secret spending; it is the act of hiding your purchases or expenditures from your spouse. I know couples who have done these types of things—kept paychecks hidden or lied about income, then spent it on themselves. Big no-no! It is, in fact, a form of cheating. The lying is a vicious cycle that can cause all kinds of problems, such as bouncing checks when the other spouse is not aware that the funds were depleted. Fees upon fees can pile up before you even know what happened!

The top cause for financial infidelity is addiction: shopping addiction, gambling addiction, porn or sex addiction, or substance abuse.[5] The spending becomes as secretive and problematic as the addiction itself. This can be so severe as to need professional counseling. If the addiction is not addressed, financial ruin may loom as the spending gets out of hand. Spouses then become so mistrustful and so devastated by the waste and shattered trust that they never recover and the marriage often ends. Marriage is worth fighting for; be honest with your partner and seek help if you have an addiction.

No tit-for-tat spending

Another problem I've witnessed in marriages is something I've dubbed "tit-for-tat spending." It can also be described as "revenge spending." I've watched couples get deeply into credit card debt over this one. One spouse, feeling entitled, makes a large, selfish purchase without spousal approval. The other spouse, resenting said purchase, decides, "Well then, I'll get something too." The original spender cannot argue, for they are, in fact, guilty of just having done the same thing. So there you go—you have now made two very large, unplanned, and unneeded purchases that have either caused debt or depleted funds that should have gone to a household expense. This quickly puts you behind and forces you to cut corners or work overtime just to catch up.

This is childish and foolish and very destructive to a relationship. If you really want something, talk about it, save for it, plan for it. Or better yet, surprise your loved one and get the desired item as a gift. This will create mutual delight in it, drawing you closer. We still talk about the time Mike surprised me with a decorative Halloween village I *really* wanted (fig. 19–1).

Fig. 19–1. My husband sold a baseball card to buy this Halloween village for me, a symbol of sacrificial giving.

It was an expensive, frivolous item, well out of budget. I could have purchased it without his knowledge or approval on a credit card, but I refrained and simply told him of my wish to have it someday, if possible. He took this to heart: later he sold one of his cherished baseball cards and used that money to buy the village as a gift—just because. Boy, was I surprised! And touched. And completely blown away. That's the kind of philosophy I'm talking about. Sacrifice. Restraint.

Chapter 19 — For Richer or Poorer

Trust. Unselfish giving. If this is your attitude, it will be reciprocated down the line for you. Things should never be put before your love for each other. B.J. Thomas sings this sentiment beautifully in one of my favorite songs: "Loving things and using people only leads to misery. Using things and loving people, that's the way it's got to be."

SHOULD MOM STAY HOME?

I suppose this is more of a question than a philosophy, but the answer to this question is the philosophy I wish to address. Should mom stay home or should she work? Research, and my own personal experience, both point to the same conclusion: women are happiest when the ultimate decision is their choice. If a woman feels compelled to work just to make ends meet when she'd rather be home with the kids, she will resent it. If she's compelled to stay home and give up a career she enjoys, she will resent it. So the bottom line is this: let mom decide if she wants to be full time, part time, or anything in between. After all, she is the one who actually has to do the work. You'll both be happier if she's given the choice. And the old saying, "If Mama ain't happy, nobody's happy" is actually true, even with the kids. When polled about what children want most, mom home or mom working, they overwhelmingly answered that they just wanted mom (and dad) to be "less stressed out and tired and happier."[6] So, whatever will make you happiest, that is what you should do; at least that's my opinion. If you're happy, the marriage will be better—that has been my experience. I would not have been happy leaving my kids in daycare; they probably would have been just fine, but I would not have been.

It is, of course, easier said than done to live on one income. Major adjustments must be made if Mom (or Dad) decides she really wants to be home with the kids. It can be done, even on a firefighter's income, and Mike and I are living proof of that. Set a budget, lower your standards of living, get rid of cable, have just one car if the children are babies (in the school years, you may need two)—these are all ways to save and all things Mike and I have actually done. We managed to put two boys through private school *and* college on one income with zero credit card debt. I have supplemented here and there, but it's mostly been on a firefighter's income that we've raised a family.

It's important to add an addendum philosophy to this category: if Mom decides to stay home, *do not demean her for not earning a wage.* A stay-at-home mom does the equivalent of five jobs. If she were paid for all that she does, some studies suggest this would be a salary worth $115,000 or more, with an estimate of working 97 hours per week.[7] Her being home saves a fortune in childcare, eating-out expenses, extra clothing, driving costs, and on and on, so believe you me, she earns her keep!

Jonathan & Denise Jones, Fire Marshal, state of South Carolina

Not long after Denise returned to work after the birth of our daughter, she came to me expressing her desire to stay at home with our children. She was very unhappy with her job, and her maternal pull to stay home with the kids exacerbated the situation. My initial reaction (although I would never have said this to her) was, "This is just hormonal. She'll get over it." I told her there was no way we could afford for her to stay home. Denise is a teacher. At the time, I was a fire department captain. She made more money than I did.

We were very fortunate that we lived in the same town as both our parents. My mom kept the kids four days a week. Her mom would come to our house to keep the kids the other day of the work week. This kept us from incurring childcare expenses and gave us people we trusted, the same caregivers who raised us, to take care of our children.

We did our best to make these arrangements work, but the desire to stay at home with the children was never far from Denise's thoughts. She was convinced that God was calling her to be a stay-at-home mom. Over and over, we would sit down with our finances to try and figure out a way to may it work. Over and over, the math never worked in her favor.

I believed that if it was God's will, he would provide a way. What I didn't believe was that God would drop her annual salary in our bank account. (He didn't, by the way.) I was waiting for God to make the math work out. (It never did.) The stresses of trying to make it work with two full-time jobs, two kids under two years old, church, and all the trimmings of life for a young couple were taking their toll.

At the time, I was assigned to the fire department's training division. Training in a combination department made up of paid and volunteer firefighters meant teaching classes during the day for the paid crews and at night for the volunteers. Denise and I were literally passing in the night. This was compounded by feelings of inadequacy, because I couldn't provide enough income for her to stay at home, and I was just about at a breaking point.

The breaking point finally came one morning when I was dropping off the kids at my mom's. I'll never forget this day. My mom asked me how Denise was doing. My reply caught me by surprise: "I don't know. She's just the lady I live with." The moment I heard those words leave my lips,

Chapter 19 — For Richer or Poorer

I knew that something had to change. Something was going to change, I didn't care what it took.

That night, I told Denise not to sign her teaching contract for next year. I didn't know how we would make ends meet, but I knew we would figure it out together. We prayed. We cried. We snuggled with the kids. We woke up the next morning with a determination and commitment to find a way to make it work. Our marriage was worth it. Our family was worth it. Our faith would be tested and so would our resolve.

It wasn't easy. We cut everything we thought we could do without: cable TV, long-distance phone service, internet, the burglar alarm system, eating out, and unnecessary trips. Still, the math didn't work. There was always more month than there was paycheck. You did not want to be around me when it was time to pay bills. If we were doing God's will, why wasn't this easier?

One day, I'd had enough of trying to make the numbers work. I was writing checks and there just want enough money in the bank to pay all the bills. I was stressed out. I told Denise to load the kids in the car and go for a ride. I needed to be alone for a while. When they left, I picked up my Bible and started pacing through the house. I was mad at God. I shouted, "I don't get it, God! We believe that we are doing what you called us to do! But why isn't this working? I'm at the end of my rope! If you've got something to say, God, say it!" I sat down on the sofa, opened my Bible, and my eyes fell squarely on Philippians 4:19: "And my God will supply all your needs according to His riches in glory in Christ Jesus." I got it. *He* will supply. Not me. If I could make the numbers work, I didn't need Him. If I could figure it out, my faith would be in me, not Him.

Soon afterward, I stumbled upon Dave Ramsey's book *Financial Peace*. This book provided financial guidance that was not just practical, but Biblical. This book helped to get Denise and I on the same page when it came to our finances. Truthfully, we still live by the financial principles of this book more than a decade later.

First, we had to figure out exactly where all our money was going. Until we did, we couldn't stop the bleeding (so to speak). Then we needed to set a budget, give every dollar an assignment, and stick with it. Money that didn't have an assignment ended up being spent on things that were frivolous, unplanned, unnecessary, and wasteful. I like to think of it in fire terminology: if you don't give every dollar an assignment, it will "freelance." Freelancing on the fireground is dangerous and often leads to crews working against each other. The same is true with your money.

In addition to having a budget, we had to figure out how to make the money we had go further. Denise began couponing. As it turned out, she was good at couponing. She wasn't into "extreme couponing" like you see on TV, but sometimes the store literally paid her to shop! Her first major shopping trip using coupons, she saved enough money to pay for our internet for three months. It became a challenge and game for her. I would come home from work and she would have all of her "goods" spread out on the counter and kitchen table. She'd ask me, "How much do you think I paid for all of this?" On average, she saved 50%–60% of our total grocery bill on each trip. We also changed the way that we planned meals. Instead of planning what we wanted to eat for the week and then making a grocery list, we learned to find what was on sale and let that dictate what we would eat.

We needed to dump our debt. If we were carrying debt, other people were making financial decisions for us. We sold our house and moved into Denise's grandparents' farmhouse, which was vacant and needed a lot of work. The floors were falling in, there were holes in the sheetrock walls, and the roof leaked. We took some of the profit from selling our house and fixed the things needed to make the farmhouse livable. We also used the proceeds from the sale to pay off our cars. We were debt-free! This freed up income that was previously going to someone else to use for our family. Freeing up this income allowed us to take vacations again. We would use my business trips as mini-vacations. We even went to Disney World for a week, and with Denise's couponing skills, we made the entire trip (park tickets, hotel, gas, food, and souvenirs) for under $2,000!

Looking back on those days, they were some of the happiest we've ever had, mainly because we were the most content we've ever been. We learned to focus on what was most important to us: our faith and each other.

CONTROLLING LEADS TO LACK OF CONTROL

Controlling is listed as being one of the top five financial reasons for a divorce.[8] Controlling can be described as one spouse delegating what the other spouse can or cannot spend. It is sometimes an actual "allotment" or allowance doled out by the controller as if their spouse were a subordinate, not an equal. "Controllers" are typically the primary bread winners, or top earners, in the relationship. This type of treatment leaves spouses feeling inferior, resentful, condescended to, even helpless.

Ironically, if you try to control your spouse in this fashion, you may so alienate them that they will flee just to get financial freedom. If this happens, it's possible they may ultimately control the money in a divorce settlement! I have witnessed this type of behavior in couples I know, and it is an ugly thing to see. Marriage is a relationship, not a job. One spouse is not the boss and the other the employee. Both are equal partners sharing the work of life and family and homemaking.

LIVE WITHIN YOUR MEANS

Debt is dangerous to marriage. Very quickly a couple can find themselves over their heads, and for firefighters, this adds a lot of stress to an already stressful profession. Avoid debt like the plague that it is. We've always viewed debt as an absolute last resort, emergency-only option.

To avoid debt, live within your means. Have a budget and stick to it religiously. This can be very helpful when money is tight and will help keep you from the emergency situations. When times have been tight for us, I clipped coupons, made very detailed grocery lists, and shopped at thrift stores and consignment stores. We bought used furniture, used cars, even used clothing and household items. Anything is better than debt, as interest compounds and you end up paying way more for something than you ought. If you live within your means, money becomes less of a crisis, and let's face it, in the world of firefighting the fewer crises arise, the better.

We all like to treat ourselves once in a while by purchasing something not necessarily in the family budget. These purchases can lead to debt and even resentment if they are frivolous or not wanted by the other spouse. Mike and I have found a solution to the self-indulgent spending problem: have a "jar." Each of us has a stash of extra cash that we save up for the frivolous but fun items we wish for. Whenever we earn extra money, a portion of that money is set aside for our jars. Mike likes to blow his fun money on anything rock-and-roll; I like designer purses and fancy froufrous for the house. There is not only zero debt or resentment with this system, as the money has been carefully set aside for such purposes, but there is mutual delight as well. I'm happy for him when he gets what he wants, and he's happy for me. And the biggest treat of all is when we contribute to the *other person's* jar. Talk about delight! Sweet and generous gestures such as these add to feeling romanced, which is especially important to me, and turns expenses from problematic to pleasurable.

You are one flesh, one love, soul mates. Your money should be shared as all things are shared in a loving relationship. The goal of marriage is to enjoy each other, support each other, and meet each other's needs. Treasure your spouse as your greatest investment, as a loving relationship is more valuable than any amount of money. Only true love ever completely satisfies. As Matthew 6:21

states, "For where your treasure is, there your heart will be also." Money should be viewed as simply a means to an end, the end being a happy home where your heart can be found.

A BIT OF MARITAL PHILOSOPHY

Before adding a few bits of what I hope will be solid and thoughtful advice on the issue of money, I'd like to set the stage with a bit of philosophy. Remember that Anne and I are not working toward a relationship that is just OK. Or manageable. Or fine. We're not writing all this stuff just to help couples live in a partnership as "friendly strangers" or in a marriage of convenience. Our desire is for you to have an incredible marriage that is mutually fulfilling, stimulating, and authentic. One that grows and builds throughout your lifetime with the ups and the downs and the triumphs and the tragedies. I say again, we are advocating for you to aim high.

SHOW ME THE MONEY (AND THE TIME)

With that in mind, I hope I don't cause you to stop reading when I say this; your priorities in life are made clear by how you spend your time and your money. There are other smaller factors to be sure, but you can "follow the money (and time)" to a fairly clear answer of what matters most to you.

We encourage you to choose your marriage and family as that which gets priority with your time and with your treasure. To assess this, you will need to do some very hard introspection and some difficult evaluation of what your stated priorities are versus your actual ones. The simple fact of the matter is that you cannot have it all. Where you spend your limited time, emotions, and energy will determine what you value. You simply cannot place equal value on everything. Some things in life get left out when we feature other things. I hope this is common sense.

Anne and I determined from day one that our marriage was going to be top priority in our lives. Our faith in God was first, then we promised to value each other above everything else. That did not mean other things weren't important, such as work, friends, hobbies, and so on. But we did decide that some things would lose when they competed with time for each other and our continued growth as a couple. This is where money starts to be a problem and why so many relationships are miserable, even destroyed, when the necessity of earning a living becomes, instead, a pursuit of wealth.

To be clear, we do not believe that there is one right way to do things. There are going to be factors in your relationship that are very different from ours. I offer up our decisions on what would get our time not as the only correct way to prioritize, but as an example of what sticking to your decided priorities looks like. This is the way we did it based on our highest choices. To keep it simple, I'll focus primarily on our relationship and family. Our basic priorities were as follows:

- We wanted consistent, quality time with each other.
- We agreed to seek what was best for each other, not what was best for ourselves.
- We committed to raising our kids ourselves.

The list is not extensive, but as priorities go, they presented significant challenges and made for some very difficult decisions. And here is where the topic of money raises its ugly head. When we committed to the above ideas, some things had to be given up. Some of those things would be people. Many of our friends wanted way more time than we could reasonably give them, especially considering the time necessary to work, raise our kids, and honor other family commitments. Not everyone had the same priorities we did and some simply did not understand why we couldn't hang out with them more or leave the kids with a sitter more frequently. Hobbies took a beating as well. Church softball, choir, firefighter teams, and other very cool activities did not always fit with what was best for marriage. Those things are fine in doses but can also begin to dominate and leave little time for each other. The biggest one, though, was work. And it is a contentious issue to talk about, but must be addressed and done so in a direct manner.

If wealth and financial security is your number-one priority, this really doesn't apply to you. You have established what you value most and thus, both of you should work and earn as much money as you can. If, as we decided, your marriage and family are the first priority, then decisions on work and earning money become more difficult. Anne and I recognized early on that in order to have the time for each other we really wanted and not have someone else raise our kids, she would stay home and create an awesome home life for our family. She did this and we do not regret any part of that decision. The consequences, however, are as you'd imagine. One income made us very tight financially, limited what we could purchase, and had a big impact on our saving for retirement. For Anne, it also came with the modern-day stigma that is attached to just being a housewife or househusband. Anne was a straight-A student and earned nearly a full ride to the most prestigious university in Washington state. She could have easily worked in the legal or academic world, and the compensation would have changed our financial situation dramatically. Doing so, however, would have also put a big damper on all three of our stated priorities. She and I both working full time would have made our priorities look more like this:

- We attempt to find time for each other around very different work schedules.

- We seek what is best for each other, but have very limited time and energy to do so.
- Someone else will be raising our kids for most of the week.

These three modifications to our priorities were unacceptable and, thus, something had to go. There is no other way to say it. Something wins and something loses. We paid the price financially to have that which we valued most: the best chance at a thriving marriage. The tight times were not always easy and we did miss out on some cool adventures that we simply could not afford. Our kids did not always get to do what their buddies were able to do, and Anne spent humbling times in the thrift stores buying used clothes instead of new ones from places like Nordstrom. We will never fully recover the money that was lost for retirement savings by not having a second income. Something wins and something loses. But our priorities were honored, and we are both eternally grateful for that.

YA GOTTA EAT

You must decide how you are going to earn enough and still be able to honor what you value most. Don't use our priorities, use yours. But understand this: you cannot have it all. If you both decide to work, your time with each other will be much more limited. If you work a lot of overtime, fatigue will play a significant role in your relationship. If you take on multiple jobs, you will have to work very hard to make time for your kids. I'm not saying it cannot be done. I'm asking you to evaluate what you are doing in contrast to your stated priorities. Do they match up?

Money is necessary and we do need to earn enough to live. But it is such a pale substitute for the glorious satisfaction of deepest intimacy with your spouse. Doing good work is honorable and necessary in life. But letting work dominate and take away from fully enjoying your family is something you'll come to regret. Toys are awesome. They are less so when they must be divided between a divorcing couple who grew apart. Prestige and respect from a successful career is good. But it is not better than the love and respect of your children.

We agreed to stand by each other for richer or poorer. Anne and I have yet to feel the richer part, but ya never know. We have spent many nights wondering how we're going to pay our bills or what gets cut out of an already strained budget. Those are the realities of the choices we made to stay true to our values and to what mattered most. When you determine what comes first, money takes its proper place—supporting your life instead of becoming it.

NOTES

1. Shellie Warren, "10 Most Common Reasons for Divorce," Marriage.com, January 31, 2018, https://www.marriage.com/advice/divorce/10-most-common-reasons-for-divorce/.

2. Larry Burkett, "Financial Smarts in Marriage," interview by James Dobson, *Dr. James Dobson's Family Talk*, http://drjamesdobson.org/blogs/dr-dobson-blog/dr-dobson-blog/2016/07/29/financial-smarts-in-marriage.

3. Dave Ramsey, "And Two Become One: How to Combine Bank Accounts," DaveRamsey.com, https://www.daveramsey.com/blog/married-couples-how-to-combine-bank-accounts.

4. G. K. Chesterton, *All Things Considered* (New York: John Lane Company, 1909), 36.

5. Douglas Weiss, *The Ten-Minute Marriage Principle: Quick, Daily Steps for Refreshing Your Relationship* (New York: FaithWords, 2007).

6. Marcus Buckingham, Find Your Strongest Life: What the Happiest and Most Successful Women Do Differently (Nashville: Thomas Nelson, 2009), 228.

7. Jenna Goudreau, "Why Stay-at-Home Moms Should Earn a $115,000 Salary," *Forbes*, May 2, 2011, https://www.forbes.com/sites/jennagoudreau/2011/05/02/why-stay-at-home-moms-should-earn-a-115000-salary/#982bc6775f40.

8. Guest Contributor, "Top 5 Ways That Money Problems Lead to Divorce," A Successful Woman, November 11, 2012, http://www.asuccessfulwoman.com/aswblog/top-5-ways-money-problems-lead-divorce.

PART VII

THE TWO TOUGHEST ASPECTS OF FIREFIGHTING

FIREFIGHTERS AND CANCER

I recently picked up Seattle Fire's newsletter, the *Third Rail*, to see what was in it. I always glance through the pictures in hopes of spotting a few faces I recognize—especially my husband's. Typically I avoid looking too closely at the section for International Association of Fire Fighters (IAFF) line-of-duty deaths (LODDs)—as do most firefighter spouses—but this day, for some reason, I was irresistibly drawn to it. Risking the tears that inevitably follow, I began to read the names and causes of death. A surprising pattern started to emerge: out of the 19 firefighter deaths listed that occurred within just 75 days, only 2 were from injuries sustained while fighting fires. The 17 remaining LODDs were attributed to the following causes: 2 from stroke, 3 from known heart attacks with an additional 4 possible (collapses following a shift), and 8 from cancer, including liver, bladder, colon, and leukemia. *Cancer is the leading cause of death for firefighters today*. It's a reality I wish I could ignore, but the statistics are just too alarming to do so. My husband has devoted much of his career to air management as a result of these cancer rates, and through him I too have become grudgingly aware of them. Cancer is scary stuff but I think *every* firefighter spouse needs to know the stats, and the earlier in your firefighter's career, the better.

WHY CANCER?

Unique to firefighters is their work environment. The dangers at a fire scene are obvious: fire, smoke, building collapse, explosions, and so forth. Firefighters can see and study these dangers and thus prepare for them as they have for decades. There is, however, an even deadlier, invisible threat emerging to take center stage—toxins.

Today's fires involve *plastics* which are now in everything—flooring, furniture, TVs, appliances, bottles, and even pipes and other building materials. When plastics burn, they typically produce much more smoke and heat than comparable wood products. But worst of all, plastic smoke is deadlier because of the gases it produces. These may include carbon monoxide, formaldehyde, nitrogen oxides, ammonia, phenol, benzene, hydrogen chloride, hydrochloric acid, methane, and the deadliest gas of all—hydrogen cyanide. If breathed in large enough doses, these gases can lead to immediate death, and even in the smallest of doses, increase the risk of cancer.

Numerous studies on firefighter cancers have been done. Notable to us, as Mike used it in his book *Air Management for the Fire Service*, is a study done by an actual firefighter. After being diagnosed with brain cancer, firefighter Mark Noble devoted his remaining time and energy to compiling statistics and cancer rates for firefighters before tragically succumbing to the disease. His courageous work was born from a powerful desire to keep firefighters safe. He challenged the fire service to take a closer look at how firefighters are putting themselves and their comrades at risk. Mark's in-depth research produced the following list of cancers that firefighters are developing at accelerated rates. Note that all four of the cancers mentioned in the introduction are on the list:[1]

- Brain cancer
- Leukemia
- Non-Hodgkin's lymphoma
- Multiple myeloma
- Bladder cancer
- Kidney cancer
- Prostate cancer
- Testicular cancer
- Colon cancer
- Liver cancer
- Skin cancer

A more recent study of firefighter cancers was done by the National Institute for Occupational Safety and Health (NIOSH), led by Robert Daniels, PhD CHP, in conjunction with the Centers for Disease Control (CDC). His research determined the following statistics; Firefighters overall have a 9% higher risk of being *diagnosed* with cancer and a 14% higher risk of *dying* from cancer than the general population. These cancers include respiratory (lung, mesothelioma) and GI (oral cavity, esophageal, large intestine, and kidney). There are significantly higher rates for some specific types of cancer, or 100% (which means 2 times the rate) of getting mesothelioma—a very rare type of cancer—and a 129% risk of dying from mesothelioma.[2] A LeMasters study done in 2006 found a twofold risk for testicular cancer and a 62% higher risk of getting esophageal cancer with a 39% chance of dying from it.[3] The overall greater risks of firefighters getting cancer than the general population as determined by the NIOSH are as follows:

- Testicular 2.02 times greater risk
- Mesothelioma 2 times greater risk
- Multiple myeloma 1.53 times the risk
- Non-Hodgkin's lymphoma 1.51 times the risk
- Skin cancer 1.39 times the risk
- Malignant melanoma 1.31 times the risk
- Brain cancer 1.31 times the risk
- Prostate cancer 1.28 times the risk
- Colon cancer 1.21 times the risk
- Leukemia 1.14 times the risk

Some fire departments are addressing occupational cancer rates that are higher than the national averages. FCSN instructors cite insurer statistics for Miami-Dade Fire Rescue: 32% (nearly one-third) of MDFR's active members were diagnosed with cancer or receiving cancer treatment between 2008 and 2010. The bottom line is this: firefighters do have a higher cancer risk than the general population, and their risks are significantly higher for some specific types of cancer.[4] Without a doubt, cancer is an *occupational disease*. And the most shocking thing of all to consider is this—firefighters are *healthier and more fit* than the general population; their rates of cancer should be *lower*, not higher.

So why are healthy, fit firefighters getting cancer? Breathing toxins in smoke is the primary culprit for these cancer rates, but skin absorption can also play a role. A study funded by the University of Cincinnati states: "We believe there's a direct correlation between the chemical exposures firefighters experience on the job and their increased risk for cancer."[5] A more recent study was done by the University of Ottawa. They took urine samples and skin wipes of firefighters before and after a fire. Their results showed an increase in polycyclic aromatic hydrocarbons (PAH) which cause DNA mutations and cancer. The amounts of PAHs were 3–5 times higher in the urine and on the skin after the fire than before the fire, which means 4 times the potential for DNA damage and cancer.[6] These studies clearly indicate the need for enhanced protective equipment, in addition to the gear already provided, that would help prevent inhalation and skin exposure to known carcinogens.

Simply stated, what firefighters breathe or touch at a fire scene puts them at risk for developing cancer. Combine the hazardous, toxic gases in smoke with the immunosuppressant effects of cortisol, plus circadian disruption from the 24-hour shift, and your firefighter becomes a prime target for disease.

CANCER: THE GOOD NEWS

It has been clearly demonstrated that firefighters are exposed to carcinogenic toxins and are falling victim at higher rates than the general population. That's the bad news, the really bad news. Is there any good news on this topic? Yes, there is! Because the cancers are toxin related they can be avoided, and we spouses can help. So what do we do? The first step is to understand the nature of toxins since they are the primary culprit in firefighter cancers.

Toxins kill normal, healthy cells before their time. This forces your body to replace those cells. With a huge influx of toxins, massive amounts of cells are killed, forcing your body to rapidly increase production. We all know that sloppy work can result when it's done too quickly, and that is exactly how cancer cells are made—from an error resulting in an over-worked system. Once the error is in place, and the body replicates the error, cancer can result.[7] This is why toxins are so deadly and why they must either be *avoided* or *quickly removed*.

Cancer is a big, scary, overwhelming topic that's hard to face. It's much easier to ignore the ugly truth and just hope it never happens to your firefighter. But I believe knowledge is power. It gives you more than just hope, it gives you tools to prevent that which is *not* simply inevitable. Nobody knows for sure why some people get cancer and others don't, but we *do* know these two things: toxins play a huge role, and firefighters are especially prone to certain types of cancer. This gives us a place to start, to narrow the topic down to manageable portions. There are very specific steps firefighters can take at work *and* at home to help prevent getting cancer down the line.

AN OUNCE OF PREVENTION IS WORTH A POUND OF CURE

The best way to deal with toxins is not to breathe them in the first place; therefore, firefighters must better manage their air consumption rates at a fire scene. Mike's book *Air Management for the Fire Service* (which is coauthored by Phil Jose, Casey Phillips, and Steve Bernocco) is the first and only one out there on this topic. These four men have dedicated their lives to preventing the number-one killer of firefighters—smoke—from taking more victims. Their book details ways individual firefighters as well as full companies can better monitor their air supplies during a fire. As my husband and his cronies have clearly demonstrated through years of research, smoke is now (because of the increased presence of plastics) much too deadly to risk breathing at all; therefore, firefighters must place the same sense of urgency on air levels as a scuba diver. This includes always leaving the fire before they use up their emergency reserves in case there is an actual emergency on the way out and they need that air. These methods have been adopted as the national standard for the fire service.

Spouses, ask your firefighter if they are wearing their SCBA while the fire is raging and even when it's just smoldering during overhaul, as the toxins are still present. Where there's smoke, there's toxins. Tell them, "I would like to grow old with you, so *please manage your air*." (My husband has affectionately dubbed his emergency reserves "Annie's Air.") If your firefighter has no idea what you're talking about, you may wish to buy them the book!

Skin absorption

Skin absorption also plays a role in firefighter cancers, therefore bunkers should be washed after every fire. Volunteers, never let your children ride in the car with unwashed bunkers! Put them in the trunk (the bunkers, not the children).

And wash those hoods! My husband told me he used to only wash his hood once it was too rank and stiff to pull over his head. Now he washes it consistently, as the hood covers the throat. Thyroid cancer is on the rise, and this may be why. Studies have shown that washing the gear with soap and water reduces the PAHs by 85%.[8] It is also suggested that dirty bunkers not be worn after a fire in the rigs or thrown onto the seats inside the close quarters of the cabin. Putting them elsewhere on the rig is something to consider on that ride back to the station.

John & Jeanne Norman, FDNY (retired)

My advice was taught to me as a probie in Engine 290/Ladder 103 by one of the senior men, Charlie Anderson. We had had a very busy day tour and gotten back from a job about 7 p.m. The night shift (which started at 6) had their meal almost ready and asked if anyone wanted to eat with them.

I remember it smelled great and I was hungry after working pretty hard and was about to say yes, when Charlie looked over and said, "Hey kid, aren't you married?" I said yes, and he asked, "How long?"

"Almost four years," I replied. "

"Does she cook?" Charlie asked.

"Yeah, pretty much every night, but I'll be late for dinner tonight with the late job and all." I said.

Charlie said, "Son, if you want to *stay* married, eat whatever she puts on the plate in front of you. She went through the trouble to make it, you go through the trouble of getting home and eating it. You'll have plenty of meals in the firehouse."

I took his advice. It was pretty tempting to sit there and listen to the postfire banter, and have a good meal with friends, but my *best* friend was at home, probably worrying why I was late, having cooked for me after working all day herself. The least I could do was get home to her and eat what she had prepared. (We had already agreed she should not wait to eat herself, since there was no telling when or how long I might get delayed.)

A key part of that lesson that I did not come to understand until much later is that our family does worry about us when we go out the door to the firehouse. They know bad things happen. They do their best to control their fears and hide them from us, but it is there, inside them, all the time. When we leave for work, most firefighters are happy as can be. We are going to a get-together with our closest friends. We are invincible. The family knows better. Share your meal and your time with them.

Nutrition

Avoiding toxins with air management and clean protective gear is the first step in preventing cancer; eliminating toxins is the second. The following are vitamins, minerals, and foods that contain cancer-fighting properties. This is a vast, overwhelming, ever-changing topic, but I've narrowed it down to a few essentials that are particular to firefighter cancers. And please please please, do your own research as well.

Vitamin D. Vitamin D strengthens the functioning of the body's "proofreader gene."[9] The proofreader gene discovers and kills errors in cell reproduction *before* they can replicate; thus, it is our first line of defense against cancer. The body needs 400 to 600 International Units (IU) a day if under age 70 and 800 IU if over 70, to produce healthy "proofreader" genes.[10] The very best source of vitamin D is about 15 minutes of sunshine three or more days a week. (After 15 minutes, sunscreen should be applied.) For those of you who live in less sunny places like Seattle, eggs, milk, and supplements are the next best sources.

Lycopene. Lycopene is a very powerful antioxidant that may help lower the risk of prostate, skin, breast, lung, stomach, oral cavity, colon and esophageal cancers. It also may improve heart health. Tomato products are very high in Lycopene (which is a carotenoid.) Carotenoids attach themselves to free radicals, or error cells, and this makes them easier for the body to eliminate. Lycopene in tomatoes is much more easily used by the body when cooked with fats such as olive oil or nuts, or is processed into tomato sauce or paste.[11] Recommended amounts vary, but adding tomato sauce to your diet is a great idea.

Selenium. Selenium is a trace mineral needed by the body. Its best sources are garlic, fish and nuts. 100 micrograms a day may reduce cancer rates by as much as 50%. (Do not exceed 1000 micrograms a day, this may cause selenium poisoning.) Selenium in the correct amounts strengthens the sanitation system in our body that removes toxins we breathe, eat, or absorb through the skin.[12]

Cruciferous vegetables. Cruciferous vegetables (which include broccoli, Brussels sprouts, cabbage, cauliflower, bok choy and kale) are rich in several different carotenoids, vitamins, folate, minerals, and fiber. They also contain compounds (indoles and isothiocyanates) that some studies show help prevent development of cancer (including breast, bladder, colon, liver, lung, and stomach) in rats and mice.[13] The recommended intake for adults is between 1.5 and 2 cups of dark green vegetables and between 4 and 5 cups of "other" vegetables (the two categories of vegetable that cruciferous vegetables fall into) a week.[14]

Nuts and berries. Nuts and berries are consistently ranked as the two highest food sources of antioxidants. One study states that people who ate nuts daily "were 20% less likely to have died during the [30-year] course of the study than those who avoided nuts."[15] Another study indicates that eating berries regularly may have potential for helping decrease the risk of chronic diseases.[16] All nuts and berries are good, but blueberries, raspberries, goji berries, strawberries, bilberries, acai berries, cranberries, and grapes (yes, grapes count as a berry) are the best, and almonds, pistachios, walnuts, cashews, pecans, macadamia nuts, Brazil nuts, and hazelnuts are the best nuts.[17] These are small, easily snackable foods. We keep a dish of mixed nuts on our counter at all times and set out a container of berries almost daily.

Other Antioxidant Vitamins. *Vitamin C.* Laboratory studies have shown that "treatment with high-dose vitamin C [vitamin C delivered intravenously] slowed the growth and spread of prostate, pancreatic, liver, colon, malignant mesothelioma, neuroblastoma, and other types of cancer cells."[18] Though prophylactic (preventive) use of vitamin C (200 mg/day) does not appear to reduce the risk of getting a cold for most people, it may somewhat reduce the length of the cold (8% in adults and 14% in children) and the severity of the symptoms.[19] Good food sources of Vitamin C are raw red and green peppers, citrus fruits and juices, broccoli, and tomatoes. *Vitamin E* also has been shown to reduce some cancers. It is recommended to take 400 IU of vitamin E daily (with vitamin C). Good sources of vitamin E include wheat germ, nuts, and vegetable oils.

Fiber. Fiber helps add bulk to the stool and moves food more quickly through the digestive systems, helping to reduce digestive cancer risks. Great sources of fiber include whole grains, seeds, oats, and beans.[20]

Testing

There are a few cancers that can be tested for, and early detection saves lives. For the general population, these tests are typically not recommended until age 50, but for the firefighter they should be done earlier. Most firefighters will resist these procedures as they are invasive and unpleasant, but it is the spouse's job to force them to go! I always tell Mike to "do it for me," and I make the appointments myself! Melanoma (skin cancer) should be checked for constantly by examining birthmarks and moles, since it's something you can do at home using the ABCDE rule:[21]

>**A**symmetry (one part of a mole or birthmark does not match)
>
>**B**order (the edges are irregular)
>
>**C**olor (the color changes throughout, featuring colors such as brown, black, pink, red, white, or blue)
>
>**D**iameter (the spot is more than a quarter inch across)

Evolving (the size, color, and/or shape of the area is changing)

Any concerning moles or birthmarks should be brought to the attention of your doctor, who can then decide whether to perform a biopsy on the area.

Prostate cancer testing requires a rectal exam. I suggest reading "Prostate Cancer: Fighting the 'Fire' Within" by John Gillis for an excellent in-depth look at how to fight and possibly prevent this disease.[22] John (a fellow SFD member and buddy of Mike's) was diagnosed with prostate cancer but is still alive and well and enjoying his retirement.

Colon cancer can be detected and prevented with regular colonoscopies. Female firefighters should get mammograms. Remember, these are some of the cancers they're prone to, so get regular tests, just in case! Have this twofold plan in place to prevent cancer: *avoid* toxins at work with protective, safe practices, and *eliminate* toxins from your body at home with a diet rich in antioxidants. And remember the importance of *sleep* and *exercise* to overall health and reduction in the negative effects of cortisol and circadian rhythm disruption as mentioned in previous chapters. Both cortisol and sleep deprivation have a negative impact on the immune system. These are all tools in the toolbox to be used to fight the chances of cancer, so please use them. Firefighter couples already risk so much for the fire service, enduring a career's worth of dangerous sacrifice for the safety of others. This is enough: don't give them your golden years too.

CANCER SUCKS

Talking about this awful thing is a drag. Cancer is like that terrible monster in horror films that just seems to be everywhere and doesn't die and just keeps coming. It's the thing we dare not speak of lest it turn its gaze upon us. At least it feels that way sometimes.

In fact, cancer is simply a reality of being alive. And for those who are married to firefighters, the odds go up that you'll deal with it in some form or another. This isn't meant to be a "Debbie Downer" type of thing, just the honest truth. As Anne so vividly detailed, the numbers illustrate that the things we get exposed to on the job elevate our risk. Even if you are fortunate enough to avoid cancer, you'll probably not escape watching colleagues go through it.

CHALLENGES OF THE FIREFIGHTER MARRIAGE

CANCER HITS HOME

I have many terrible experiences with this and it feels like a new one pops up every week. The deepest hit in my professional life came when my dear friend Chief Jesse Youngs was traveling to Arizona to see the home he ultimately hoped to retire to once done with SFD. While getting on the plane he felt a paralysis on one side of his face, a harbinger of the hellish descent into brain cancer that was to come. This is still painful for me to recall. Jesse was a buddy, a mentor, and one of the finest men I knew. He was an exemplary fire chief and a great leader who was loved by the members of our department.

Cancer certainly didn't care that Jesse was only a few years from retirement and had just gotten his dream assignment as deputy chief of operations for SFD. Cancer had no concern whatsoever that he had at last met the girl of his dreams, had a grandson who was the light of his life, and had just purchased a hard-earned retirement home in the sun. That he was in good physical shape, had love, and was as deserving of a good ending to his career and a happy retirement as anyone I have ever known didn't matter a bit to the big C. It took my dear friend in a matter of months and left us all with a gaping, awful hole in our lives. There are many other stories like this and most firefighters have them. This one still grieves me and never seems to fade (fig. 20–1).

Fig. 20–1. Deputy Chief Jesse Youngs with Mike at a training fire. Courtesy of Lt. Tim Dungan, Seattle FD.

NAGGING IS NOT ALWAYS BAD (IT USUALLY IS)… BUT NOT THIS TIME

We realize that the fire service is certainly not the only profession to have high cancer rates. But we are one of the highest and it isn't going away. As Anne has carefully listed, there are some things you can do to knock down the odds that this devil will come knocking. Do all you can to minimize exposure and strengthen resistance. Consider her suggestions and do some research of your own. I'd like to appeal to spouses to be on your firefighters like a dog on a bone to do the right things. This is an area where it's OK to nag and guilt trip them to take precautions. I'll highlight two that Anne mentioned that are particularly critical and that spouses won't see, as they happen at work:

Don't breathe smoke. Remind them to wear their air mask and to breathe as little smoke as possible. It is pure poison and will disorient, choke, and asphyxiate your firefighter. It also contains massive amounts of carcinogens. We advocate that you remind your firefighter that the air in their mask is their lifeline. And there is a small part of that air, called the emergency reserve, that is intended to get them home to you and your family. In our classes, we called that "Mama's Air," because that air gets you home to Mama in an emergency.

Wash your skin and your gear. After every fire make sure to shower thoroughly and get the firefighter bunking gear into the washing machine. Current studies are showing that cancer-causing chemicals such as PAHs (polycyclic aromatic hydrocarbons) readily cling to our gear and are easily transmitted through the skin.[23] Tell your loved one to wash and wash and wash some more. Anne and I spent a little extra on our current bathtub to get a feature called Microsilk that may help with toxin removal. It shoots millions of tiny bubbles into the water that collect on the skin and pop. This process seems to show great success in helping people with skin disease. It is also used in burn units as it promotes healing. Its use for toxin removal has yet to be determined, but we like our odds that it may be effective.

To the firefighter reading this, you owe it to your family to do everything in your power to keep cancer at bay. Yes, it's a drag to always wear your mask. Yes, it's a pain to strip down your gear and get it in the washing machine, especially for those of you with departments that don't give you two sets. Figure it out. Find a way. Your family needs you.

THE FAMILY THAT HELPS TOGETHER

One last comment regarding cancer and firefighting. Not all of us will have to deal with this nightmare, but every one of us will know those who do. It has been amazing to watch other firefighters and their spouses rally around the families of those who are in the midst of this terrible fight. I am particularly proud of the many instances Seattle firefighters have stood side by side with families whose firefighters are sick. Truly epic stuff. And the help is critical because quite often the city will deny cancer coverage. The state will often deny that the cancer is related to the fire service even when the cancers are covered by presumptive legislation. Families who are already staggered by a tragic diagnosis are not always in the frame of mind to deal with this stuff.

A hallmark of great marriages is standing together to do good for your fellow firefighters in need. As a team, you can truly help others who are devastated, scared, and unsure of what comes next. It may be simply listening or sharing a good cry. It could be providing meals or contributing to expenses. Or helping to connect the dots with the very confusing medical and legal paperwork that is necessary but difficult, especially when grieving. Whatever the needs, they can be done together with strength and caring. You may be able to include your kids as part of the team, when appropriate, and build in them a sense of community and compassion. We've talked at length about the destructiveness of selfishness. Being there for those who are hurting in your second family reflects all that is good in humanity.

YOU HAVE TODAY

A final thought in this bummer of a chapter. Most of you reading this don't have cancer right now. Your firefighter is healthy and you can do pretty much anything you want. Pause for a moment and reflect on that reality. If sickness comes your way you will have to figure it out, but it's not here right now. What is here now is the wonderful opportunity to love each other, enjoy your life, and be there for those who are less fortunate. So dump all the minor garbage that clutters up life and make it awesome. We obviously can't guarantee that cancer won't be a part of our marriage. What we can do is work to minimize the odds, comfort those who are sick, and strengthen our bond as one flesh. Should it someday be us who must face this horrible disease, being a solid couple will be imperative to navigating our way through. There is no stronger team than a husband and wife who are committed to each other and to making the world around them a better place. In sickness, and in health.

NOTES

1. Mike Gagliano, Casey Phillips, Phillip Jose and Steve Bernocco, *Air Management for the Fire Service* (Tulsa, OK, PennWell, 2008).
2. Robert Daniels, Travis L Kubale, James H Yiin, Matthew M Dahm, Thomas R Hales, Dalsu Baris, Shelia H Zahm, James J Beaumont, Kathleen M Waters, Lynne E Pinkerton, "Mortality and Cancer Incidence in a Pooled Cohort of US Firefighters from San Francisco, Chicago, and Philadelphia (1950–2009)," *Occupational & Environmental Medicine*, vol. 71-6, BMJ Journals, June 2014, 388–397.
3. University of Cincinnati, "Firefighters Face Increased Risk for Certain Cancers," *ScienceDaily*, November 10, 2006, https://www.sciencedaily.com/releases/2006/11/061110080741.htm.
4. "Firefighter Cancer Fact Check," Firefighter Cancer Support Network, http://www.firefighterclosecalls.com/wp-content/uploads/2017/06/FF-Cancer-Fact-Sheet.pdf.
5. University of Cincinnati, "Firefighters Face Increased Risk."
6. Jules Blais, "Firefighters Absorb Harmful Chemicals through Skin, Study Finds," University of Ottawa Department of Biology newsletter, October 18, 2017, https://science.uottawa.ca/biology/news/firefighters-absorb-harmful-chemicals-through-skin-study-finds.
7. Michael F. Roizen and Mehmet C. Oz, *YOU: The Owner's Manual* (New York: William Morrow, 2005), 332.
8. K. W. Fent, B. Alexander, J. Roberts, S. Robertson, C. Toennis, D. Sammons, S. Bertke, S. Kerber, D. Smith, and G. Horn, "Contamination of Firefighter Personal Protective Equipment and Skin and the Effectiveness of Decontamination Procedures," *Journal of Occupational and Environmental Hygiene*, October 2017, http://www.tandfonline.com/doi/full/10.1080/15459624.2017.1334904
9. Roizen and Oz, *YOU*.
10. National Cancer Institute, "Vitamin D and Cancer Prevention," October 21, 2013, https://www.cancer.gov/about-cancer/causes-prevention/risk/diet/vitamin-d-fact-sheet.
11. "How Lycopene Helps Protect against Cancer," Physicians Committee for Responsible Medicine, http://www.pcrm.org/health/cancer-resources/diet-cancer/nutrition/how-lycopene-helps-protect-against-cancer.
12. Roizen and Oz, *YOU,* 337.
13. National Cancer Institute, "Cruciferous Vegetables and Cancer Prevention," June 7, 2012, https://www.cancer.gov/about-cancer/causes-prevention/risk/diet/cruciferous-vegetables-fact-sheet.
14. MyPlate, "All about the Vegetable Group," Department of Agriculture, updated November 3, 2017, https://www.choosemyplate.gov/vegetables.
15. Julie Corliss, "Eating Nuts Linked to Healthier, Longer Life," *Harvard Health Blog*, November 20, 2013, https://www.health.harvard.edu/blog/eating-nuts-linked-to-healthier-longer-life-201311206893.

16. Baoru Yang and Maaria Kortesniemi, "Clinical Evidence on Potential Health Benefits of Berries," *Current Opinion in Food Science* 2 (2015): 36–42, https://doi.org/10.1016/j.cofs.2015.01.002.

17. Ruairi Robertson, "The 8 Healthiest Berries You Can Eat," *Authority Nutrition* (blog), *Healthline*, July 31, 2017, https://www.healthline.com/nutrition/8-healthy-berries.; and Robertson, "The Top 9 Nuts to Eat for Better Health," *Authority Nutrition* (blog), *Healthline*, December 5, 2016, https://www.healthline.com/nutrition/9-healthy-nuts.

18. National Cancer Institute, "High-Dose Vitamin C (PDQ®)—Patient Version," updated April 20, 2017, https://www.cancer.gov/about-cancer/treatment/cam/patient/vitamin-c-pdq#section/all.

19. Office of Dietary Supplements, "Vitamin C," National Institutes of Health, updated February 11, 2016, https://ods.od.nih.gov/factsheets/VitaminC-HealthProfessional/.

20. "Food and Cancer Prevention," ASCO Cancer.Net, September 2017, https://www.cancer.net/navigating-cancer-care/prevention-and-healthy-living/food-and-cancer-prevention.

21. Stacy Simon, "How to Spot Skin Cancer," American Cancer Society, July 6, 2017, https://www.cancer.org/latest-news/how-to-spot-skin-cancer.html.

22. Jon Gillis, "Prostate Cancer: Fighting the 'Fire' Within," *Fire Engineering* 160, no. 12 (2007), http://www.fireengineering.com/articles/print/volume-160/issue-12/features/prostate-cancer-fighting-the-ldquofirerdquo-within.html.

23. Fent et al., "Contamination of Firefighter Personal Protective Equipment."

Firefighting Is a Deadly Calling

It's shocking how life can suddenly change from familial bliss to horrific spectacle with a simple phone call. A call that brought numbness, a heart-pounding chill, a dull ringing in the ears. Somehow, we knew that our lives would never be the same. Mike's profession up to that point had been all cool stuff—fire trucks and hoses and ladders. He'd spoken at our kids' school and church groups in full bunker gear—pretty exciting for little boys to display Dad in his regalia to their classmates and church mates. Dad was a hero—his job was fun and he got to dress differently from the other dads. Dalmatians and horns and flashing lights were tools of his trade. Daring rescues and gushing water hoses—all in a day's work. And who could forget the fire pole? Every kid loves to slide down one of those. But firefighters possibly dying, burning to death, why would we ever consider this grisly reality? We as a young firefighter family had never had to deal with the more dangerous aspects of Mike's job. But when the phone rang that night long ago, it was a grim wake-up call to inform us that firefighting is a deadly calling.

It's January 5, 1995, and I am hosting the seventh birthday party for our youngest son Rick. His actual birthday is January 1st (he was the first baby of the new year born to Rapides Parish, Louisiana, in 1988) but since New Year's is a holiday that most families choose to celebrate together, we decide to do his "kid party" on the fifth instead.

One of the biggest dilemmas a mother faces when planning her little boy's birthday party is where to have it. Six- and seven-year-old boys are quite rowdy and destructive, especially en masse. I already know this from having planned a party for our eldest son, Michael, just the year before. (Our boys are only 19 months apart in age.) Michael's birthday is in late May, however, so his rambunctious buddies could be corralled outside in the yard where there is little to be damaged. But I can't do this for Rick's gang, as his birthday is in the winter. So, what to do? I have two choices; either face the destruction of the interior of my house or risk the humiliation of out-of-control little boys in a public place. I opt for humiliation over destruction.

Near our home there is a brand-new McDonald's with a big indoor play area. Again, it's 1995, and these indoor playrooms are something new. It has giant slides and tubes and various crawl spaces, much like a large hamster cage, only for small humans. But the most coveted and appealing aspect of this new place is the ball pit. This cage-like area boasts an indoor trampoline covered waist deep (for short people, that is) in multicolored plastic balls—cutting edge stuff for 1995. I decide to host Rick's party there. He's thrilled at the prospect (fig. 21–1)!

Fig. 21–1. Celebrating Rick's seventh birthday on January 5, 1995.

We arrive at the new McDonald's around 4:00 p.m., along with 15 little guests and a few brave parents. Most parents, however, choose to bail and will return only when forced to. Our little group takes up an entire section of the restaurant. The place is decorated with balloons and colorful birthday napkins and cups. I add party favors and gift bags and a homemade cake. The kids all get funny hats and Happy Meals. It is perfect; it is chaotic, noisy, and frantic beyond description, but wonderful at the same time. The stress of trying to keep up with running, scattering, crazy little boys for whom I am responsible threatens to be overwhelming. But I must keep things in perspective or I will miss the joy of the moment, which is watching my little boy have a perfect seventh birthday.

As our little party begins near our home, north of Seattle, something else begins just south of the downtown part of the city. On South Dearborn Street in Seattle's commercial district, a member of a local band shows up for practice. He notices smoke coming out of the Mary Pang Chinese Food Company's warehouse. He calls 911. The Seattle Fire Department rolls up to this one-alarm fire and quickly takes action. The warehouse has several levels and multiple entrance points. Though this type of structure is typically quite dangerous, it appears to be a very small fire on the lowest level, so the crew is not overly alarmed. They quickly contain the visible fire on the ground floor and are confident it is nearly out. They are wrong, dead wrong. Unbeknownst to them, for they have no floorplan of the building, this warehouse has a basement that is fully engulfed in flames assisted by accelerants raging undetected beneath where they stand.

I'm frustrated; I can't get Rick out of the ball pit. I can't get *any* of the kids out. They roll and jump and bury themselves in the colorful plastic orbs, unconcerned with my schedule. It's time to eat and open presents—it's getting late—we must move along. But I mutter to myself again to keep it all in perspective, they're just little kids; don't let agitation ruin the joy of the moment. I relax, smile, and take pictures of my little boy buried in plastic balls.

At the warehouse, the fire is not contained. They've called for more crews; it is no longer a one-alarm fire. Eleven firefighters are on the ground floor, baffled by the high heat and heavy smoke with no visible source. Then suddenly the source reveals itself as the floor gives way. An illegal modification, or "pony wall," has burned through causing a portion of the floor to collapse beneath them into the basement. Six of the eleven firefighters manage to jump out windows and doors; one clings to the remaining fragments of the missing floor and pulls himself to safety. But four aren't so lucky; they plunge helplessly into the raging inferno below. All hell breaks loose. What began as a one-alarm fire will become a five-alarm fire that will burn on through the long night.

I finally get Rick and his buddies out of the human hamster cage with its funky ball pit. He opens his presents—he cheers with delight and gratitude for each one. He poses with his firefighter dad as I take pictures. The kids all talk (or yell) at once. Their exuberant, high-pitched voices hurt my ears. They spill their drinks. They get frosting on their clothes and in their hair. They pop the balloons. I'm getting a headache. We've gone over our allotted party time, and I am now getting the evil eye from McDonalds employees. The world of firefighting is the last thing on my mind, though Mike, even now, wears a firefighter sweatshirt. As I pack up gifts and return rowdy boys safely to their parents, I again try to keep it all in perspective. Though exhausting, little boys truly are joyful creatures, and I've somehow successfully completed yet another little boy party. Mike and I smile at each other over the heads of our children—life is good, we love being parents, we love being together. If I could give young firefighters and their families only one piece of advice it would be this: slow down, enjoy the sweet moments and soak them up to the fullest. Cherish each other every single day, for it can all change on a dime. It did for us and for other firefighter families that night of our young son's birthday party. We were, in that moment, blissfully unaware of what would come crashing down around us when we later pulled into our driveway. Our perspective of the fire department was about to drastically change forever.

When we got home, our neighbor and fellow Seattle firefighter Tom Nelson called as soon as he saw us pull in. He'd been watching for us, anxious to tell Mike of a major fire—the Mary Pang warehouse fire—that was ongoing. Tom had heard that there were reports of firefighter injuries. He was barely out of recruit school, so Mike, the five-year vet, casually replied that there were always rumors of injuries at fires. They usually ended up being smoke inhalation or sprained backs. No big deal—nothing to be alarmed about.

Since Mike seemed unconcerned, I was unconcerned as well. We got the kids inside and I quickly got their sticky, party-grubby little bodies into a bath in preparation for bed. We were happy, we were at peace, then the phone rang again. Mike had been called in off-duty before, but tonight would be a very different event. The Mary Pang warehouse fire had gone to five-alarms, was raging out of control, and worst of all—there were four firefighters caught in the collapse of the main floor, trapped somewhere in the hellish inferno of the warehouse's basement.

Mike grabbed his uniform and was immediately out the door. It was an abrupt and unsettling turn of events to what had been a wonderful day. He briefly told me what was happening prior to leaving, and my blood turned cold. This was my worst nightmare, every spouse's worst nightmare, hearing of their firefighter being trapped in a fire. Were they hurt? Were they suffering? I couldn't even let myself imagine the final possibility. He had to go and try to help those already on scene do everything they possibly could to save these guys; they were his brothers, even though we did not as yet know exactly who they were.

Mike raced to the scene; it was a hellish one. Crews were working feverishly to extinguish the raging blaze as the building was coming down around them. Everyone there was willing to risk all if there was any hope of getting those men out of that fire. The horror of what Mike saw that night still haunts him today. He set about his tasks with a new level of urgency never before experienced.

We didn't have cell phones back in those days, so I was in for a long, sleepless night. The boys were finally asleep after repeated questions as to why Daddy had to leave so abruptly. Rick's birthday presents sat piled on the dining table, already all but forgotten. The joy of this special day was now replaced by their confusion and my anxiety.

I sat glued to the TV screen hoping for bits and pieces of news on the fire. When I finally got a glimpse of it, my heart sank and my eyes filled with tears. The blazing inferno that was once a building still trapped the four firefighters, and I knew their chances of survival were pretty slim. But maybe their gear

Chapter 21 — Firefighting Is a Deadly Calling

would protect them, or perhaps they had found a safe niche, or better yet—they could have already escaped unbeknownst to the reporters. I had to hope these things for them and for me, because without hope, I wasn't going to survive this night, let alone the many years ahead as a fire wife. The scene was surrounded by firefighters and I knew my husband was there, risking everything to try to save his friends.

It was a long night. I waited and waited for him to return, to hear word, to see if the fire was out, but no word came. Firefighter spouses over the years have felt this heavy weight of tension, of not knowing, of nearly crawling out of your skin with anticipation as your loved one is out there, somewhere, battling death itself. I had not known this feeling before, but now I did, and it was terrible. I suddenly wished with all my heart that my husband was anything but a firefighter. That he could be home safe in bed with me, and I cried myself to sleep with tears of self-pity and resentment and fear.

When I awoke the next morning, Mike was still not there. I was nearly sick with disappointment. I somehow got my little boys off to school, then returned to my hellish vigil. Finally, his car pulled in—I nearly screamed with relief at the sight of him. He never looked so good to me, despite the fact that he was haggard and black with soot, his eyes red-rimmed and bleary. He seemed to have aged overnight. When he saw me, he grabbed me in a desperate embrace and his tears began to flow.

"They're gone," he croaked, "We couldn't get to them in time. My brothers are dead."

Keep it all in perspective, I thought in that moment, be strong for Mike, don't break down. Four families have just been destroyed; my heart is broken for them. But my family is okay, for my love has safely returned to me this day. That is the juggle of this life—holding on to hope and joy so as not to be crushed by grief and worry and helplessness.

Lt. Walter Kilgore, age 45; Lt. Greg Shoemaker, age 43; Firefighter Randy Terlicker, age 35; and Firefighter James Brown, age 25 (fig. 21–2), entered into eternity the same night we celebrated our son's birthday party. For us, now, the two events have been forever linked. We could let the sorrow of this tragedy ruin our son's birthdays, but we choose instead to keep it all in perspective; we choose to remember the fallen four with love and fondness and respect, and we are grateful that Rick's birthday *never* lets us forget them.

Fig. 21–2. In memory of Lt. Walter Kilgore, Lt. Gregory Shoemaker, FF James Brown, FF Randall Terlicker.

How does one begin to describe in words what it feels like to lose four co-workers, four comrades, four friends, to a fire? All I can truly say is that it changes you. Mike, regrettably, didn't know Lt. Walter Kilgore personally, but he knew of him, as his reputation was stellar. Likewise, he had not met FF James Brown who was newer to the department. He did, however, know Lt. Greg Shoemaker. Mike had only worked with him once, just once, but so instant was their mutual respect for one another that Shoemaker mentioned the possibility of an opening at his station and wondered if Mike was interested. Mike considered it. Now Shoemaker was gone.

We both, however, knew FF Randy Terlicker—knew him quite well. "Terly-Bird" was in Mike's recruit class. Anyone who'd ever met Randy never forgot him. Randy was larger than life; he was not only a giant, burly bear of a man, but he was boisterous, jolly—you know the kind—he simply lit up a room with his presence. At Mike's recruit class graduation BBQ, Randy had grabbed me in a big hug as if he'd known me my whole life and said, "So this is Mikey's wife!" He was in my heart, he was in everyone's heart, from the very first moment I met him. Lovable, likable, and fun—that was Randy. And Randy loved being a firefighter; he was so into it that he'd stage his bunkers and boots very carefully to gear up on the way to the rig, instead of just putting it all on at the door. When asked why he did this he'd say, "So I have even more time to get them on correctly before I get to the rig—I'll be that much faster." His enthusiasm was legendary, and so was his appetite. He'd clean out the station fridge of any and all leftovers, no matter how old they were. On his off-days, he'd visit his classmates at their various stations just to stay in touch. Mike loved him—everyone did.

Over the next two days, Mike went down to the fire scene to join in the gut wrenching search for Randy's body through the mountainous, charred piles of debris. Randy was not with the other three who had already been found. Mike was not there on the third day when at last they located him. As his body was carried out on a stretcher, The Mary Pang warehouse fire claimed its fifth victim. FF Gary Medica, Randy's cousin and close friend, was so distraught that he went into full cardiac arrest. He was revived on scene but was forced into early retirement as a result of the damage done to his heart. Less than two years later, Gary would succumb to a final, fatal heart attack.

Randy had always said that if his brothers or sisters ever fell in a fire, he would remove his own PASS device, leave it with them, and go get help. He was teased for being so dramatic. But apparently this is exactly what he had done; it is believed that this is why his body was found away from the other three, without his PASS device, which was discovered amongst the others. Randy, the big-hearted, bear of a man, had somehow survived the catastrophic fall into the basement and summoned enough strength to crawl some distance away in his last, desperate efforts to save them, leaving behind the very device that was meant to have saved *him*.

How do you describe the funerals of such men? Again, all I can say is that they changed us. Randy's recruit class wanted to do something special for his funeral. Mike offered to write a poem; the class then had it inscribed on a plaque which they presented to the family. The family loved Mike's poem so much that they had it inscribed on Randy's tombstone. Having my husband's words etched in stone in this way is humbling and haunting, a painful honor.

> *A giant man, a mischievous grin*
>
> *A loving son, a faithful friend*
>
> *A tender heart, a caring brother*
>
> *A man of honor, like no other*
>
> *with strength and courage, soft yet tough*
>
> *A hero—but even hero does not say enough*
>
> — *Mike Gagliano*
> (on behalf of Seattle Fire Dept. Recruit Class #55)

As a result of the Mary Pang warehouse fire, the Seattle Fire Department has been changed forever. Fire officials now tell the crews affected of arson threats; before that did not consistently happen. Communication has drastically improved, as has protective gear—from hoods to thermal imaging cameras (TICs). Buildings are now routinely inspected, and building plans are kept on a computer database as well—as all of these deaths could have been avoided if they had but known there was a basement in that warehouse.

Seattle firefighters are now given more self-rescue training, and rapid intervention groups were created to get to fallen firefighters more quickly. But the biggest change of all was the establishment of a training facility. One of the settlements from the fallout of lawsuits required that Seattle at last have a proper training facility within the city itself. It took 10 years to build. It was not occupied until 2006—the year my husband Mike made captain.

Mike's first assignment as a captain was to the JTF (joint training facility). It was a match made in heaven. He hates it when I brag about him, but too bad—it's my prerogative—my right and privilege. Mike was an *amazing* training captain; the center's secretary called him "my wild red-headed child," because his enthusiasm and vision for training was so infectious that it energized the division. He won officer of the year as a result of his work while he held that position.

I often hear this word when people describe my husband's work for the fire department and all the extra-curricular training he does, and that word is *passion*. Mike is, indeed, a man of passion—passion for me, for our kids, for his faith, and for the fire department. He pours all that he is into all that he does—but training firefighters takes center stage above other hobbies or off-duty pursuits. Many have wondered why that is, and now I can tell you why—it's largely due to that fateful night many years ago—the night four friends died. The night fellow classmate Randy died. It has left an indelible mark on Mike's heart. Casey Philips, fellow "Seattle Guy" and coauthor of *Air Management for the Fire Service*, was also Randy's recruit classmate, and he too, picked up the mantle of training alongside Mike as a result.

And it has changed me forever too. I no longer sit idly by, oblivious to what my husband faces when he heads to work. Never again can I delude myself that he'll safely return, as he may not. I have faced a sobering reality—the reality that a monster lurks beneath the surface of every call, a monster that may choose to strike.

Around the same time of the Mary Pang warehouse fire I saw the movie, *Backdraft*. This was a tough film for a firefighter's wife to watch because firefighters die and there's a funeral with marching uniformed firefighters and bagpipes. It is far too real, I know, because I've been to them. But I was confronted with something in this movie that I had not really thought about. The main character, a stellar firefighter played by Kurt Russell, is separated from his wife played by Rebecca DeMornay. He tries to reconcile, as there is clearly still love there. But she refuses, shaking her head as she says, "I can't, I just can't... the chances you take..."

This puzzled me—she didn't want to be with him because *he might die?* Had she deliberately chosen to harden her heart because she wasn't willing to risk losing it? I suppose the option of self-preservation is one way to deal with a deadly calling. But I've chosen a different path.

With risk comes reward, with sacrifice comes blessing, and there is no greater profession on earth than saving lives. Firefighting is indeed a brother/sisterhood with bonds that go deep, painfully deep. But the joy of this profession, the pride and honor of it all, far exceeds the anxiety, and the threat of death makes life all that much more precious and worth fighting for. I believe the four fallen heroes would agree.

Friends and family outside the firefighter world have asked me if it's hard to live with the constant threat of death. Is it depressing? Do I resent it? I know this profession can kill my beloved at any time. Firefighters have one of the highest on-the-job mortality rates in the country. I've had my share of panic attacks; if I'm talking to my husband on the phone and the alarm bell goes off in the background, my heart races, even after all these years. It's not easy, and it's not pleasant. But I've chosen to make it a positive. Life is short; therefore, it is to be cherished.

Some see the stress of this job as a negative, as a wedge that can drive couples apart, but we choose to see it as a very powerful reason to stay close. Each day, each moment is a gift, so make the most of them. We tell each other every day of our love. We never take life for granted. And I choose to appreciate my husband for the hero that he is; he saves lives for a living. In my opinion, there is no calling more noble. He has my respect, and my heart. Losing my heart is a risk I'm willing to take.

Chad & Rebecca Simons, Unified Fire Authority, Utah

When Chad and I were engaged, I received a particularly unsettling phone call from him. He was working for the Salt Lake County Fire Department (now Unified Fire Authority) at the time and I was working as a graphic designer for a carpet cleaning company. As I sat at my desk making my way through the day's tasks my phone rang. I answered it and the first words I heard were, "Don't worry, I'm not hurt."

While that statement quickly communicated to me that Chad was safe and uninjured, it also brought the immediate realization that he quite easily could have been hurt. At that moment I had no idea what danger he might be in or what the situation was, other than the fact that he was at work.

I found out later that his paramedic partner had suffered a serious injury while on the scene of a fire. Here is what happened in Chad's own words:

"My partner and I were assigned to assist a ladder company with an elevated master stream operation at the scene of a multiple alarm fire. The fire had burned through the roof and the new plan was to extinguish the fire from above using ladder trucks. My partner beat me to the stick and began maneuvering the nozzle around to deliver the water needed to put out the fire. At some point during the operation several events occurred that culminated in the extension of the hydraulic ladder. My partner had very large feet that extended past the pedestals at the top of the ladder and into the rungs. The unexpected extension of the ladder caught one of his feet between extending rungs and nearly ripped his foot into two pieces. Several of us quickly rushed to provide care. This was a large fire in the middle of a busy city, and news media was covering every move. I knew that Rebecca was aware I was working in that area and would think the worst as the breaking news came across the wire announcing a firefighter being seriously injured, so I gave her a phone call as soon as I could."

Working for the fire department is inherently dangerous. The phone call I received that day could have relayed much worse news, and that's something the spouse of a firefighter always has floating around somewhere in their mind. Sometimes it remains in the back corners, and sometimes it can loom constantly at the forefront. It can be difficult to keep the fear and worry at bay.

That phone call continued to bother me for some time, and I wondered how I would cope as the wife of a firefighter, knowing the dangers of the job. As it weighed on my heart and mind I made my feelings known to God through prayer.

Chad and I were married shortly after that, and within those first few months of marriage something changed in me. I began to notice that my fear and worry had all but disappeared. I still felt it on rare occasions, but it wasn't something that weighed on me regularly. As I pondered this I had a sudden moment of clarity, and I knew that my faith and belief in God had brought me peace and taken away my fear.

That was fifteen years ago, and to this day I still feel the peace of mind it brings me, even as Chad is in potentially dangerous situations at work. People sometimes ask me how I handle being the wife of a firefighter and the fear and worry that often come with it. I still know that at any time I could get a phone call much worse than the one I received that day, but it's not something that I worry about regularly. Through my faith and through God I have found the peace and comfort I need.

LIVING LIFE ON THE EDGE

"Danger gleams like sunshine to a brave man's eyes."

—Euripides

There is no escaping the fact that fighting fire is a dangerous job. Hiding from the realities of burning buildings, twisted metal, collapsing structures, deadly disease and all the other craziness that firefighters call work is not really an option. You took each other on for better or for worse. This is some of the worse.

IT'S A FAMILY AFFAIR

And it impacts the whole family. I recall the day the planes hit the towers in New York City on 9-11 and watching them collapse. Every one of us knew that many lives were being lost at that moment and it was sobering. What I did not know was that my youngest son, then 13-year-old Rick, would also watch these events and they would have a lasting impression on him. Prior to 9-11 he was interested in what I did at work but not overly spooked by the dangers. On that day, the full reality of what could happen struck him deeply. He would demand that Anne call during the day to see that I was OK and jump when the phone rang to see if everything was all right. It was heartbreaking to see the fear in his eyes and hear it in his voice. So great was the impression left on his tender heart that it is likely he will end up doing some type of anti-terrorism work as an adult.

WORST DAY OF MY CAREER

It remains hard to think about the Pang warehouse fire and reading Anne's description was tough. The images are still crystal clear in my mind and my stomach gets queasy reliving the sights and sounds of the worst day of my fire service career. I never fully understood the true dangers of the job until that day and the reality of it spurred me to seek the best training I could find. The memories of digging through the rubble, in an effort to locate Randy, was a surreal nightmare that still makes me shudder. At one point the ATF pulled us all out as we thought we had located him. We were furious that there was to be a delay as we did not want him in that pit one moment longer. It was personal. They, of course, were right in wanting to preserve any evidence possible to help convict the murderer, Martin Pang, for the crime. At Randy's funeral all I could do was weep at the loss of such an amazing man's life and how awful it was for his family. My fellow recruit school classmates and I could only hold each other in shock.

When I responded to the fire, a veteran officer requested me to be on his crew as I had some truck experience. We were assigned to the front of the building and could see down into that hell hole of a burning pit from the doors and windows. The building sagged dangerously, and the walls bulged out. At some point, the officer had to be relieved as he worked the next day, and I was put in charge of the crew. It was sobering to be responsible for these men as the dangers from the gutted smoldering building were real. Especially given how distraught and emotionally raw we all were. One of my crew members, Jeff Birt, would later succumb to another terrible aspect of our job: duty-related cancer. Sitting at Jeff's funeral I recalled the sense of frustration and confusion we all felt at the Pang fire and being totally helpless to save our beloved brothers. The horrible irony was not lost on me that day that Jeff survived the flames of the Pang fire only to be taken by another hideous danger, the toxins from smoke.

These experiences and many, many more are simple testimonies to the realities of this dangerous calling. Anne and the boys had to face up to the very real possibility that I might not come home from work one day. That is a challenge that all firefighter spouses must recognize.

MADE FOR THIS... AND IT MATTERS

I hope you will find comfort first and foremost by recognizing what we do really matters. This is not some thrill-seeking adrenaline ride that is more about kicks than necessity. What we do impacts lives every day and contributes to a civilized society that works. Firefighters show up to provide help during the worst days of people's lives. The danger and risk are about bringing order to chaos and help to those in dire need. That should bring some sense of solace amidst the fear.

In addition, there are a few other things to keep in mind that set firefighters apart and enable us to meet these terrible challenges head on:

We work with incredible people. Our fellow firefighters are warriors and they would gladly risk their own lives to see that your spouse gets home to you.

We train for these emergencies. It is a daily commitment to honing our skills so the things that look so risky are much less so.

Chapter 21 — Firefighting Is a Deadly Calling

We expect the bad stuff to happen. Unlike the average citizen walking around, we show up anticipating bad situations. They don't catch us by surprise and we are prepared to handle them.

We have cutting-edge training and safety information. We are in a time in history where knowledge is abundant and some of the best training in the world is readily available. Many ideas and examples of how to do our job better are now on the internet or in training classes. This allows us to learn from others' mistakes so we don't make them ourselves.

We were made for this and truly enjoy helping others. Firefighters accept the calling because it's who we are. Most spouses recognize this, and it does provide comfort.

I don't mean to minimize the dangers or be cavalier about how risky some of our runs can be. But there is truth to the thought that these bad things must be taken on by someone. And that someone is us. We are proud of what we do and very good at doing it. As Franklin Delano Roosevelt so accurately said:

"Courage is not the absence of fear, but rather the assessment that something else is more important than fear."

The needs of our community, at their very worst times, are more important than the risks, the dangers or our fears. They are the environment where our courage, training and desire to help come together into often heroic feats of service. That should inspire and comfort you as a spouse. You are a part of our work and play an invaluable role in helping us to fulfill our calling.

HOME MATTERS TOO...

And, please remember, it's just fine to remind us that we are needed at home too. That you and the kids are counting on us to walk through that door at the end of a shift and be a part of the family. Risk at work is necessary and understood to be a part of being a firefighter. Unnecessary risk is not. Remind your firefighter of all that needs protecting at home.

Fighting fire is dangerous and at times deadly. All spouses who choose a firefighter as a mate will have to navigate the fears and anxieties that are a natural part of this life. Anne came across a quote from Lt. Colonel Dave Grossman, in his book *On Combat*, that helped to calm her mind. It didn't eliminate the dangers but brought what I do into focus. She will tell you it helps to consider these words: "There is something gloriously right with them. Because if we did not have warriors, men and women willing to move towards the sound of the guns, and confront evil, within the span of a generation our civilization would no longer exist."

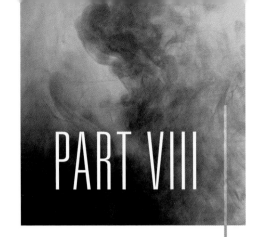

PART VIII

ENDING ON A POSITIVE NOTE

THE POWER OF ENCOURAGEMENT

We often look to nature for inspiration. For examples of beauty, grace, and the miraculous. One such image that inspires me, that touches my heart and encourages me in my efforts to have a successful relationship with my firefighter husband is this: the flying V formation of migratory birds.

In the fall I hear them as they pass overhead, the migratory birds of the north making their way south for the winter. Since we live on the water, we are directly on their route. I often rush out to see them when I hear their soulful cries, and there they are, high up in the sky forming the perfect picture of encouragement and support that is an image for us all to appreciate. You might be thinking, what in the world do migrating birds have to do with marriage, or for that matter, firefighting? Everything, as a matter of fact. Migrating birds are incredible creatures with astonishing achievements of endurance and courage and strength (as are firefighters.) The record holders (the Arctic tern) fly the equivalent of *three trips to the moon and back* over the course of a lifetime.[1] This distance would be amazing for human beings to endure, but it is absolutely astonishing for a seemingly mindless little bird to do so. How do they go so far and survive? It is because of the formation of the flying V, which is the embodiment of teamwork and encouragement in action.

The V formation allows the birds to take advantage of the aerodynamic effects of flying behind another bird. This effect is known as "aerodynamic wash-up," which creates an updraft from the flapping of the wings of the bird in front.[2] The bird flying in the "upwash" position literally gets a "free-lift." Birds in the back of the V have been shown to have lower heart rates and flap their wings less often than the bird in the lead position[3]

And what about the bird in front? There is no benefit to them and no explanation as to why they would ever willingly take this position. In fact, it is the toughest spot of all to ever be in. The lead bird meets the most resistance and tires the quickest of all, yet remarkably—every V has a "volunteer" in front, leading the way and taking the hit for the team. This bird is often the strongest, but not necessarily the youngest.[4] In fact, studies show that the older birds have the lowest mortality rate during migration, about 5%, while the juveniles have the highest rate—about 35%. Youth apparently isn't what it takes to survive, but unselfishness. The young fly all over the place and often ignore the V. They are not born with the instinct to fly in formation but must learn it. Those that learn, survive. Those that don't, perish. Migrating is tough business: starvation, dehydration, immune compromise, and the intense physical demands take their toll.

The V formation is brilliant as it drastically cuts down the mortality rates of those that apply it. With it, birds can fly 71% further than without it! They reach their destinations quicker and easier.[5] In an example of teamwork at its finest, the birds rotate the lead position and take turns leading and following. As a result, they only fly in the front about 32% of the time, and no one has to lead the whole way. Not only do they change position in the V, but they adapt wing flaps as well to achieve maximum thrust. Each bird carefully synchronizes their flapping to that of the bird in front, creating an efficient air flow pattern for each and every bird in the V.

But the flying formation is more than just position and flapping rhythms. Migratory birds such as the Sandhill crane, the northern bald ibis, and the Canadian goose also *communicate* with each other nonstop. Their honks and cries are a continual source of affirmation and encouragement to the leader as he is breaking the wind resistance for the rest. If you could put their honks into words they would sound something like this, "How ya doin' up there, Bob, ya need a break yet?" If Bob says yes, then another bird shifts forward as Bob drops to the back, and the team moves ever forward without having to stop.[6]

BE A COUPLE AMID THE FLOCK

Canadian geese provide yet another deeply moving example of support and help. When one goose falls out of formation, two other geese fall out with it. They remain with that goose to protect it from predators until it is strong enough to carry on. And the two go one step further: they provide a small flying V for the weak bird, using their own bodies to break the wind resistance till they can all rejoin the bigger "V."[7] Amazing.

Within the flying V formation of the northern bald ibis, an intimate, almost human pattern exists. The ibises actually fly in *pairs* among the flock. Every pair constantly looks out for each other, noting and monitoring the well-being of just that one bird. In this way, they are acutely aware of each other's strength and even the amount of time they've spent leading. This extra attentiveness prevents freeloaders from taking advantage and overtiring any one bird. It is a beautiful picture of marriage itself.[8]

"Though one may be overpowered, two can defend themselves" (Ecclesiastes 4:12). We are not meant to be alone. To be alone is to be more easily overpowered by trials and tribulations and foes intent upon your destruction. Two are not so easily overpowered. Why is this so? Because with two, someone has your back. With two, one can rest while the other works. With two there is support and teamwork and strength, and most importantly, encouragement.

An inspiring example of what two or more can accomplish together is displayed in the flying V formation. It is a visual symbol of what unselfish teamwork and heroic encouragement can achieve. To *encourage* is to *give courage*, to cheer on, to inspire, praise, console, reassure, strengthen, revitalize. It is one of the greatest gifts you can give to your spouse. For any marriage to be next level, to go the distance, to survive the onslaught of enemies, encouragement from one another is essential. This is particularly so for the firefighter marriage. Firefighters need courage to face the ultra-intensity of danger and trauma, and the firefighter spouse needs strength to try to meet the needs of her depleted, weary warrior. It is a draining profession for both. *One may be overpowered, two can defend themselves.*

LOVE LANGUAGES

The group is good, but no one encourages like a spouse. Words, actions, belief, touch—these have the power to diffuse stress and give energy. Encouragement should be a regular, practiced part of the firefighter marriage. And by practiced, I mean put into action and carried out on a regular, daily basis. What does your spouse need? What can you do to encourage them? In his book, *The 5 Love Languages: The Secret to Love That Lasts*, Gary Chapman asserts that there are five different ways to express love and support to your spouse. These five are *words of affirmation*, *acts of service*, *receiving gifts*, *quality time*, and *physical touch*. Everyone places different value on these five, so it's important to learn the ones of greatest importance to your spouse, and "speak" most often in these. I, for example place a much higher value on words of affirmation than Mike does; and he places a much higher value on physical touch than I do. Make a plan, make it a daily practice. Ask yourself, How can I encourage my sweetheart today? This can be accomplished in as little as ten minutes a day.

Words of affirmation

This can be something like a text, a phone call, a loving note left in the morning. Compliments, praise, gratitude. Tell them what they do well. Say how proud and grateful you are to be their spouse, and let this be known to others as well.

Thank them for all their sacrifices for the family and unselfish acts of support for you. Tell them that they give you courage, that with them you are stronger. Say that you need them. You have more power than you know. Your adoration and praise will make them; your harping and criticism will break them.

Acts of service

Do something nice for your spouse. Complete a task that would normally fall into their realm. This can be anything from vacuuming the house or making sure the dishes are done when they get home. To surprising them with dinner, to fixing the car, mowing the lawn, or picking up the kids. Take something off your weary spouse's plate. And remember that slaying dragons is exhausting work.

Receiving gifts

Surprise and delight your spouse with gifts, not just when they're expected (like holidays and birthdays and anniversaries) but all the time. This can be flowers, jewelry, a favorite snack, or an addition to a cherished collection. These gestures show your spouse that you're thinking of them, listening, and remembering their wishes and desires. My husband recently arranged for me to swim with dolphins, something I've always dreamed of doing (fig. 22–1). It was one of the most magical days of my life, and he made it happen. Find out what they delight in and buy them gifts that support these pursuits. Mike wants to learn to play the guitar, though it may seem a bit late in life to start. Instead of objecting, I bought him a Gibson Flying V guitar (a fitting symbol to this topic). Dream big Babe, I'm already your biggest fan and always will be.

Fig. 22–1. Livin' one of my dreams, thanks to my hubby

Quality time

Go on dates. Take time to be together, just the two of you, whether that's over a nice meal at a restaurant, playing fun games, or going on a hike. Make your spouse feel first. Firefighters, let them know they are first in your heart even above the major draw of the firehouse. Put down the cell phone and focus just on them, at least once a day. Say yes to their needs more often than you do to the needs of others. And firefighter spouse, find out how your firefighter likes best to spend their days off and try to make that happen.

Physical touch

Hold your beloved close and touch them in the ways they need. Whether its sexual or non-sexual, touch communicates love and support in a way that is unique to marriage. Make sure you are communicating your desires in this area clearly to each other and seek to meet those needs. A loving caress, a kiss, and making love—all soothe and heal even the most weary of firefighters and firefighter spouses alike. And touch keeps you close, intimate, and strong. It is encouragement in its physical form.

Just as migrating geese communicate in soulful honks as they fly great distances with much resistance, encourage one another in whichever love language is specific to your marriage, to your spouse. Communication binds two souls together. Together you can go further, together you are stronger, together you can withstand the challenges of life, even the challenges of the firefighter marriage. One may be overpowered, two can defend themselves.

Phil & Emma Paff, Queensland Fire and Rescue, Brisbane, Australia

To start, I'll give a number of marital points in no particular order:

1. **Firefighters, learn to say no.** If you are a committed worker that is so into your job, you'll either find things to make better or fix or develop. Or because you're engaged, your employer will continually target you to make better, or fix or develop. It consumes time and, at some point, will take you away from family. I (Phil) have developed and taught rescue courses since 1997, and my aggregate for sleeping away from home is 18 months! So for 1 year and 6 months I've lived somewhere else. I began winding this back in 2011. At the end of day, it still is only a job.

2. **Know each other.** It sounds so easy, right? The stresses of the job, yes, we all think about the bad calls, fires and accidents. Most

times a lot of angst comes from organizational crap, people in the workplace, the employer or government. For the partner, knowing when to push and talk or when to give space is an art. You need to know each other. I know that Emma is concerned about what has happened at work but will give space until I'm ready to vent about it. At the same time though I know she is there.

3. **Yes, you have two families, but...** Not so much a problem here in Brisbane but my time in FDNY showed me that some firefighters will prioritize "the guys" and going out and having fun at the expense of family, repeatedly. At the end of the day, the job and your old long-retired, moved-on fire buddies won't be the ones at your bedside when you exhale your last breath.

4. **Have your own interests.** It's okay to not spend 100% of your time off together. We all need to have our own identities. It's what attracted us to each other in the first place.

5. **An old (but useful) maxim—never let the sun go down on an argument.** Talk, talk, talk. Especially in this day and age, we have a no-phone policy at dinner/evening time.

6. **Keep it sexy,** however that may be, from date night for some to cleaning the toilet for others. Nothing kills relationships more than routine.

7. **Avoid temptation.** I don't know about fixing this one, but I see so many guys now who use the smartphone to Facebook-stalk women, Tinder, watch porn, and so on. Temptation is at the end of your fingertips and the porn can normalize what might be some pretty bizarre behaviors.

8. **Laugh.** We can be so serious at home with the demands of building extensions, mortgages, educating kids, car payments, utility payments, places to be on time, and so on. We spend a lot of time at work laughing with the boys and forget to do it at home. Laugh at home with each other. Have fun.

9. **We have different skills, but we are equals.** For many, being a firefighter is not just a career but a calling. This is a truly noble undertaking, but it can be all-consuming and dominate not only a firefighter's life but that of their family. In many ways, we define ourselves as firefighters and not as regular people who happen to be firefighters, and sometimes this can lead to feelings of superiority. Our families, our partners have jobs, careers, pursuits, and aspirations of their own. We must (truly) respect and support those aspirations, as being a tag-along partner to everything fire will likely not be their main focus. Support your partner, as they do you, in their pursuits.

Emma is a self-employed patisserie chef who creates truly extraordinary cakes for any occasion. With the training and level of commitment required it is very much a calling, like being a firefighter. My mind boggles at her creations and the level of realism and detail they contain. Some of them seem to defy gravity at times!

Emma supports everything I do as a firefighter. She not only understands the 'job' but also the 'brotherhood'. It is just as important that I support Emma with her career, and by that I mean an authentic and nonpatronizing support. Recently Emma won State Patisserie Chef of the Year; I was so proud to see all her hard work culminate in that one moment.

As a couple, nothing is one sided: she chose me as much as I chose her, we're a team. So when Emma embarked on this career path, we as a couple had to build it up, which meant literally building a professional kitchen for baking and a studio for customers, all in the backyard!

Fast-forward three years and everything's going gangbusters, so now the challenge for us is striking a balance with work and life. For us that can mean setting some small but important rules, such as no mobile phones at the dinner table or after 7 p.m. Our electronically connected society places unnecessary pressures on people to reply to anything at all hours, so we put those devices aside at night because it's our time to share a meal and talk (a dying art, it would appear).

HOW I ENCOURAGE MY FIREFIGHTER

When I think of a firefighter, the word "courage" comes to mind. If anyone truly does embody the word, it is, indeed, a firefighter. Bravely they face danger to save lives. And selflessly they lift broken bodies from carnage and wreckage as they risk their own feelings to give comfort to the hurting. They face danger and trauma, day in, day out, all year round. They exhibit more guts, more grit, and more backbone than the average citizen can possibly comprehend. I know this because I've lived with one for more than 30 years.

Mike casually tells me of his acts of heroism with hardly a hint of the dramatic. "Yeah," he'll say, "I delivered a baby today," or, "Went to a stabbing—guy had a steak knife in his neck," or "Did CPR in a grocery store, guy died on aisle 5," and "Had a house fire today, no big deal." My jaw drops in wonder or horror, my eyes bulge with emotion—and he barely raises an eyebrow. So much courage, so much strength; I stand humbly in awe.

True heroism is remarkably sober, very undramatic. It is not the urge to surpass all others at whatever cost, but the urge to serve others, at whatever cost.

—Arthur Ashe

I am fortunate that my husband's nature is quite positive. He is a high-energy, upbeat kind of guy with lots of passion and enthusiasm for life. In my travels as a speaker on firefighter marriage, I've met firefighters from across our country, Canada, and even Australia, and everywhere I go I have found that they all typically have the same nature as my firefighter's. They are confident, capable, and caring—the three Cs, as I call them. This seems to be a job requirement. They give and serve and rescue with undaunted courage and an almost cavalier, devil-may-care attitude. Amazing and fearless—I love 'em!

But I also know what a toll this job can take on a firefighter because I've witnessed it, up close and personal. Yes, my husband's nature is positive, and yes, he is inherently brave—but he does, on occasion, battle depression. It happens to him, just as it can happen to any firefighter. And it is no wonder. He sees some pretty grisly things, as they all do. Firefighters routinely clean up society's messes while society, in horror, turns away. They keep chaos from ruining us all and they do so for not the highest of wages. They give and give and give and ask very little in return.

This is where the gift of encouragement comes in. It is one that will help keep the firefighter and the firefighter marriage from going over the edge. Again, encouragement primarily means "to give courage to," and who needs it more than a firefighter? They need it more but ask for it less. As was mentioned in previous chapters, a human being can endure incredible amounts of trauma and stress if they have just one person in their life with whom to share it.

Firefighter spouse, don't ever underestimate your worth. You are vital! My firefighter loves it when I encourage him. It doesn't take much effort and it doesn't cost a thing, but it is *invaluable* to his health. In lifting him I also lift myself, as his nature is restored and he gives and gives back again. The result is lots of breathing room for the extra stresses in life and restored hope to carry on. Will he ask others for encouragement? Never. But he asks me for it—he needs it—and this I give him gladly. I feel it's the least I can do.

How exactly does one encourage another? Every relationship is a little different, but in our marriage it looks like this: I've always considered myself to be the unofficial memory keeper in our family, as I'm the one who takes the most photos. I like to remember—especially the good stuff. This is fortuitous, as my weary firefighter *loves* it when I tell him the good stuff. He'll often request this

Chapter 22 The Power of Encouragement

as we lay awake at night talking quietly in the dark. I used to wonder why he repeatedly asked me to do this, then one day I read what high stress does to the brain; it actually *shrinks* your memory.[9] My highly stressed firefighter sometimes literally forgets the good stuff. It was an Aha! moment for me.

Since it is such a regular request, I have become quite good at it. I have a list. I keep it on the ready with fresh updates. I begin with our marriage, since it is our top priority. I say how wonderful it is that we found each other and how delightful it is that we are still best friends after so many years. Next are the children. I list all their positive qualities and point out their potential. All parents spend a lot of time and emotion worrying about the kids; their needs and weaknesses can be overwhelming and even heartbreaking at times. That's why it's refreshing to remember all the good in them. Mike loves to hear that "they'll be all right…they'll make it in this world."

The house is third on the list. I point out how blessed we are to even have one. Many people in this world do not. It's cozy, clean, and comfortable. It's a home, a place in which we've made countless happy memories. A sanctuary from the storms, a place to rejuvenate. Houses, especially ours, have many flaws and need lots of upkeep. Focusing on the positive reminds us both to be grateful and content and to not worry so much about the next big fix.

I point out all that is good with our extended family, country, health, achievements, and our faith. Our bodies aren't perfect—we know we need to lose weight and exercise, always—but sometimes it's nice to just be grateful in the moment that our hearts are still beating and our lungs are still breathing. Families always have issues, but I remind Mike of the joy we've had with them all. I point out that we still live in a free country—one that we believe is the best the world has ever seen. And I remind him that God is in charge, not him, so relax.

Then last, but not least, I tell him how proud I am of him, how awesome his job is, and how much I believe in his calling to be a firefighter. People all over this country try and try to get this job and fail, but not my husband! We locals have this saying: "Getting hired by Seattle Fire is harder than getting into Harvard," and you know what? It's statistically true.[10] His career is exciting, challenging, rewarding, and vital. Sure, he doesn't get paid as much as say a banker or a business owner—but his role in society is invaluable. Without him, others couldn't do what they do! And he loves his job while others despise theirs. I admire and respect him and tell him so with all my heart. He gets more satisfaction from my appreciation than from paychecks; most firefighters do.

Encouragement brings out potential and sparks genius. For both men and women, honest admiration from a spouse motivates and rewards. It staves off depression and feelings of hopelessness that can occur when dealing with life and death. Without emotional support, a firefighter's confidence, which is a major source of his or her success, erodes and eventually crumbles. People who only ever receive criticism from their spouses will likely have less confidence. Be your beloved's most enthusiastic fan.

Encourage each other—*give courage;* cheer on, inspire, praise, console, reassure, strengthen, and revitalize. To receive this much-needed encouragement, ask yourselves these two questions: Are you flying in V formation, firefighter couple? And are you a supportive pair within the bigger V, watching each other directly while staying close to the bigger firefighter family for extra support when needed? Together you will go the distance in this high-stakes game of life and death, of service and sacrifice, but only if you do not fly alone. Mortality rates soar for the inexperienced who fly solo. Take turns out front and take a break when needed. Always communicate, both to cheer and to ascertain need. And be a couple amid the flock: know each other better than anyone else and never let freeloaders take advantage to the point of overtaxing your endurance. Stay close to the flock, and stay close as a pair by knowing your preferred languages of love. And tell each other the good stuff.

GOOD MEDICINE

Encouragement is energy. It is magical, mystical stuff that gets a bad rap as mushy or silly, and that's too bad. In the world of the firefighter, it goes against a culture that rightly does not seek its own accolade. The work and service is the priority and when that is done well, the reward is felt deeply and it is enough. That's the way most of us feel and it comes from a good place, an honorable place. As it turns out, it is typically not sufficient. Oh sure, it works for a while, even for years for you tough old birds out there. But life is a challenge and there are plenty of things that will drag you right down to the dirt. Health, family, work, politics, religion, sports (if you're a Mariners fan), neighbors, friends—the list goes on. We all need a friend. This is true of your spouse and critically true if your spouse ingests chaos, carnage, and despair as a daily diet.

DOWN BUT NOT OUT

It is embarrassing to read Anne's words that mention I get down or depressed at times. My goal is to be strong and inspiring and tough. I don't want others around me, and particularly my fellow firefighters, to be bogged down by anything negative I'm feeling. But I do get glum and sometimes it's beyond what I can get a hold of. There is often not a specific reason I can point to or an answer as to why it is that a funk has settled in. But it happens, and Anne knows it immediately. I hope this is because it's not all that common and so stands out from the norm. However it comes about, the majority of the time it is she who pulls me out of the tailspin. Typically, it is the advice she has just given you, played out in real time, that does the trick and keeps the wolves at bay. And I am so grateful. My goal is to always be there to return the favor.

THE GRATITUDE ATTITUDE

One of the key things I believe couples should foster in their relationship is to approach life, both as individuals and as a team, with a grateful heart. Gratitude simply overwhelms so many of the negative emotions that seem to plague folks today. I know when I am writing up invitations to my particular "pity party" of the moment, some honest recognition of all that I have that is good cancels the party. And my guess is that every one of you has so much to be thankful for that your celebrations of misery will vanish too.

I don't mean to undermine the reality of sadness and depression. There truly are some cases that need both medical and psychological intervention. If that's you, please get professional help. But most of us are just allowing the difficulties of life to cloud out the good stuff. Speaking personally, we have had some brutal things come down the pike in the past six or seven years, including terrible deaths in our family, losing beloved pets, loss of friends, professional setbacks, a flooded basement, and other things you can likely relate to. These events naturally created terrible emotions that were difficult to handle amid trying times. But they are only a part of the story. If you truly want to keep the dark clouds in perspective, what is also true for our last six or seven years is as follows:

- We had our first grandchild (Hallelujah chorus sounds).
- Neither Anne nor I had any serious medical issues (we've had minor ones though).
- We never missed a meal (and ate a few too many, actually).
- My job is solid and enjoyable (even during an economic downturn).
- Our house is warm and cozy.
- Our kids are healthy.
- We live in the United States of America (talk about winning the lottery).
- We actually have a savings account and an IRA.
- We get to teach and speak to wonderful people all over the world.
- We remain happily married after over 30 years together.

We got to see Kiss, Queen, The Eagles, Aerosmith, Sammy Hagar, Van Halen, Cheap Trick, Def Leppard, Styx, Boston, Foreigner, Great White, Joan Jett, Lita Ford, Guns N' Roses, Neil Diamond, Halestorm, Pat Benatar, Creedence Clearwater Revival, the Beach Boys, and way more cool bands than I can remember in concert.

- We've taken numerous Disneyland and Disney World trips.
- We watched the Seahawks win a Super Bowl (on TV, but still).
- We attended our son's graduation from law school.

This list could be pages and pages of good, uplifting things. And they are as much the truth as the bad stuff that drags at our hearts. An "attitude of gratitude" is a decision you make that is not founded in self-help trickery. It is an acceptance of truth—real truth. A good deal of the encouragement you can give to your spouse is simply reminding them of what is positive and awesome and right in your world. Believe me when I tell you it will help—and help in a very big way.

BUT WHAT ABOUT...?

There are a few obstacles to encouragement that should be considered, as they are the primary reasons this great medicine is not often administered:

- Why should I give encouragement, I need it too
- Too focused on the unimportant matters of life to see it is needed
- Feel it's silly and will be laughed at
- Too proud to accept encouragement
- Don't know the right thing to say or do
- Too stinking selfish and lazy
- Too drunk, drugged, or overmedicated to think clearly

Once again, lots more could be listed, but I'd simply encourage you to look at these and quit doing them. Don't think about quitting, just stop. Every one of these has its source in selfishness and a decided lack of appreciation for all the blessings you've been given in life. Encouragement is a simple, loving, unselfish, and inexpensive way to help your spouse navigate challenges. It will draw you together and increase your trust in each other.

SHARE THE LOVE

Some additional notes that should be added to the mix. The first is that it must be reciprocal. Anne describes this well and she is dead on. It's important that you get it through your head that what your spouse needs in the way of encouragement may very well be different than what you need. Sex is not the answer to all that ails (though that tends to be my first choice when it comes to staving off depression). Anne often needs very different forms of encouragement and it is my job to figure those out and do them, not just do what I want under the phony guise of "encouragement."

PERPETUAL "DEBBIE DOWNER"

Second, if you are the one who is constantly down and constantly needs encouragement, maybe it's time to do a gut check. Are you approaching life with a determination to focus on what you do have, as opposed to all the ways life has done you wrong? If it's the latter, you are going to be a miserable mess and a drag to be around and live with. It is the responsibility of every spouse to build up their mate and encourage them in life. But it is also everyone's responsibility to carry their own water and figure out how to be happy.

HOW YOU TALK MATTERS... A LOT!

Finally, this idea of being encouraging should also dominate the way you speak to and about your spouse. The fire service culture can be a bit harsh and dripping with sarcasm. So are many of the occupational environments of spouses. Familiarity can breed a biting form of contempt because we know each other so well. It is an absolute necessity that you get a hold of your tongue when it comes to speaking to your spouse, especially in times when they may be weak. Your words will be a weapon instead of a comfort. This is never more important than when you are with others. Nothing is more devastating to a partner than being mocked, diminished, or criticized in front of others. It is embarrassing, humiliating, and destroys trust. It also typically results in a tit-for-tat bashing that is the complete opposite of all we've discussed in this chapter. I have mentioned selfishness as the primary destroyer of relationships and that is true. But one aspect of selfishness that is immediately evident is the disparaging way some spouses talk to and about each other. When we witness this in the couples with whom we associate it always rings a warning bell. It is the sign of a relationship in trouble, and usually does not end well.

OF GUITARS AND GEESE AND...

The imagery of the geese taking turns, encouraging, and protecting, is so inspiring. I hope the imagery of Anne getting me a Flying V guitar, despite my total lack of musical aptitude, brings a smile. It doesn't have to make sense all the time, it just helps. The look on her face when she gets a little love note or a gift she wasn't expecting is priceless. Encouragement matters. Light words of cheer work like nothing else to dispel the shadowy dark clouds. The wolves are out there and they come with teeth bared and in a wide variety of disguises. Standing together, side by side, you have the strength to stand. And to enjoy the ride whatever trials may come your way.

NOTES

1. Shanta Barley, "Arctic Tern Crowned 'King of Communters,'" *New Scientist*, January 13, 2010, https://www.newscientist.com/article/dn18379-arctic-tern-crowned-king-of-commuters/.
2. Robbie Gonzalez, "Why Do Birds Fly in a V Formation?," Gizmodo, January 17, 2014, https://io9.gizmodo.com/why-do-birds-fly-in-a-v-formation-1503746314.
3. George Dvorsky, "Why Birds Take Turns at the Front When Flying in V-formation," Gizmodo, February 5, 2015, https://io9.gizmodo.com/why-birds-take-turns-at-the-front-when-flying-in-v-form-1683465294.
4. Ibid.
5. Robert N. Lussier and David C. Kimball, *Applied Sport Management Skills*, 2nd ed. (Champaign, IL: Human Kinetics, October 21, 2013).
6. Ibid.
7. Ibid.
8. News Staff, "Northern Bald Ibises Take Turns When Flying in V-Shaped Formation," Sci-News.com, February 3, 2015, http://www.sci-news.com/biology/science-northern-bald-ibises-geronticus-eremita-v-shaped-formation-02458.html.
9. Deane Alban, "12 Effects of Chronic Stress on Your Brain," Be Brain Fit, https://bebrainfit.com/effects-chronic-stress-brain/.
10. David Brewster, "Seattle Fire Department: Harder Than Harvard to Get Into," Crosscut, October 29, 2007, http://crosscut.com/2007/10/seattle-fire-department-harder-than-harvard-get-in/.

23

Divorce Is Not the End

I *hate* divorce. I hate what it represents. The destruction of something special, divinely ordained (in our humble opinion), and meant to last forever. I feel so strongly about the sanctity of marriage that I have devoted many years of my life working to help save it, especially among my favorite folks. And obviously, my work has led to this book—the culmination of mine and Mike's efforts in this area—all with the hope of helping firefighters save their marriages, their families, and ultimately, their lives.

Divorce is never victimless, and unfortunately the ultimate victims tend to be children. As I wrote earlier, it has a life-altering impact on kids. For this reason alone, divorce is something to be avoided as much as possible. Do everything you can to save the marriage as it is *almost always* the better option. I believe that marriage is the most intimate relationship for us to aspire to, a union of body and soul. When it ends, it is tragic, painful, devastating. Marriage is worth fighting for with everything you've got.

It is my opinion, which is derived from years of study and experience with this topic, that many divorces in this country occur for very ordinary reasons. Typical reasons include communication issues, money, growing apart, or being married too young. There are certainly terrible situations of abuse and infidelity, but most divorces we have encountered have been for very fixable reasons. Our hope is to provide some help to see that "ordinary" problems plague us all, and many of those can be rectified. And the belief that all problems will end if "I can only find someone better" is often just a terrible, *terrible* lie.

The truth is this: statistically speaking, second marriages (resulting from divorce) fail more often than first ones do (and third marriages even more so).[1] The pressures of combined families coupled with financial woes can make the second marriage even more complicated and stressful than the first one. In trying to rid yourself of old issues you simply take on new, potentially more difficult ones. And sadly, in seeking a "perfect" love or perhaps a more exciting one, you may lose something irreplaceable: a close relationship with your children.

In trying to make your life better, you've only made it worse. Another reason why I hate divorce.

And this is the reason I hate divorce most of all—because of the destruction it can render upon the disillusioned and the brokenhearted. The following is a very tough subject to broach, but one that is critical. It may not seem appropriate in a book about marriage, but the two are, in fact, directly connected. There is a shocking new monster creeping its way into the fire world and that monster is *suicide*. These suicides are linked to depression and anxiety. Depression from firefighting? Perhaps, but the primary culprit may in fact be *depression from divorce*. Men who are struggling financially are 10 times more likely to commit suicide than affluent ones.[2] Divorce cuts an income in half. A married firefighter is much less likely to ever develop PTSD; a *divorcing* firefighter is the most likely candidate of all.[3] Under the influence of extreme stress (like that experienced during a divorce), one can make very poor choices. The continuous traumatic stress of the job on top of the stress of divorce (along with subsequent financial repercussions and separation from beloved children) can make the firefighter—one who *saves* lives for a living—extremely vulnerable to the poorest choice of all; to end their own life.

Chapter 23 — Divorce Is Not the End

I'm writing this on the cusp of a recent, heartbreaking tragedy: a firefighter's suicide within our own personal sphere. This suicide stemmed directly from a bitter divorce. Suicide is an alarming trend for firefighters nationwide, and it is breaking our hearts. "A firefighter suicide strikes at the very core value of the profession—teamwork."[4] When the fire department loses a brother or sister, the loss is felt by all. And losing them to suicide is especially tough, as a once vibrant member of a life-saving team has become isolated and alone, falling into the utmost level of despair.

The Centers for Disease Control and Prevention (CDC) reports the following statistics on suicide in the United States:[5]

- There were 44,000 suicides in 2015, which breaks down to an average of around 121 a day.
- Suicide is the 10th leading cause of death overall.
- Men have suicide rates approximately four times higher than women.
- Each suicide affects at least 18 people
- 70% of all suicides are white males
- For women, individuals employed in protective service occupations (police, paramedics, firefighters, etc.) have the highest suicide rates (14.1/100,000).[6]
- Chicago FD had 7 suicides in 18 months from the beginning of 2007 to the middle of 2008. This triggered a study of member deaths from 1990 to 2010, which found 41 suicides, all men, with an average age of 55.[7]
- Risk factors for suicide include biopsychosocial factors (depression, anxiety disorder, family history of suicide, childhood abuse or trauma, etc.), environmental factors (social loss, such as *loss of a significant relationship*; access to lethal means; being a victim of harassment, etc.), and sociocultural factors (difficulty seeking help, no access to mental health care, exposure to suicide or suicidal behavior, etc.).[8]
- Divorce is the number-one trigger for suicide.[9]

A recent study on the correlation between divorce and suicide done by the National Institute for Healthcare Research in Rockville, MD, stated the following: "Divorce now ranks as the number-one factor linked with suicide in major US cities, ranking above all other physical, financial, and psychological factors."[10] I believe that divorce, thus, is a major issue for the firefighter as they are already at risk for suicide because of the stress of the job. Add to this a bitter divorce that may bring financial and legal struggles, and the threat is through the roof. It is an unfortunate reality of this profession, one that the firefighter brother/sisterhood should be aware of.

I've stated quite clearly all the reason I hate divorce and all the reasons I believe it should be avoided. But sometimes it is unavoidable. Sometimes, even after your greatest and noblest efforts, you may still find yourself getting divorced. Firefighters, if this should be you, please remember that while it is indeed the sad end to a chapter, it is not the end of your story. As bad as it is, divorce is *not the end*.

> ### Rob & Angela Hughes, Baltimore County FD, Massachusetts
>
> We started our relationship in the fire service in November 1992 as recruits with the Baltimore County FD. We lived close to one another and decided to carpool to work. We kept each other in check, made sure we didn't forget anything, and became the best of friends. We always trusted one another in good times and bad. But it wasn't until January 2001, after we both went through divorces, that our friendship blossomed into something more.
>
> We both had young children, so many people told us our relationship would not work. The stress of the job and maintaining a combined family were difficult. But there was a chemistry to our relationship that we knew was like no other, and we were committed to exploring the future.
>
> It was not just our lives that were affected but also the lives of our young children, who ranged in age from two to eight. This was a critical, difficult time for them, and they were our priority. After trying several churches, we found one that welcomed us and our blended family. Immediately we felt comfortable with the pastor and knew this was the start of something special. As our relationship moved forward, we decided officially to become family in March 2003. The first step our pastor recommended was making the kids a part of the ceremony. He explained that it was just not the two of us getting married; we all were.
>
> He had the most wonderful idea of taking vows with our children. During the ceremony, we presented each of the four children with a pendant of three circles. The circles represented the past, present, and future and the blending of the six of us into one unit. In our vows we promised to love the children, care for them, and help them to be all that they could be, listen to them, talk to them, and forgive them. Most important, we vowed to stick together and have many good times as a family. We made our children a priority in our marriage.
>
> Having time for one another, without the kids, is equally important. We agreed that a part of our past failures involved failure to set aside time to be a couple. We came up with date night, a night committed to just us, a time to unwind and a time for intimacy. When money was tight, we

Chapter 23 — Divorce Is Not the End

> ordered carry-out and watched television. We promised each other a night without interruption. We agreed no one would cook or answer the phone. And above all we agreed not to let the kids intrude. We looked forward to this time, and even now, with our children grown, we still have date nights.
>
> We have based our marriage on respect, trust, and, most of all, communication. We have pledged to always talk to one another, never to walk away from one another during a fight, and never to go to bed angry.
>
> As fire department members, we do not recommend working together. The toughest part of our fire service relationship occurred recently when we were both assigned to the same station, on different shifts, as captains. This was no ordinary firehouse; it was the special operations station. We commanded an engine, tower, two medics, an EMS supervisor, and the urban search-and-rescue (SAR) unit. Staffing and operations changed constantly. Sometimes we did not agree. We did not want to bring work home, but it was hard not to. One night at shift change, we did not agree about the staffing changes. Our disagreement continued into a heated phone conversation. The station rookie came into the watch room and said, "Capt—Umm, the other Captain Hughes is on the phone and does not sound happy." Later the troops said we were like "Mom and Dad fighting," and we knew it was time for a change. We are now assigned to different stations.
>
> With all that said, what we do is an important part of who we are. We confide in one another, understand the stress of the job, and rely on each other. Sharing the love of the fire service with your life partner is a beautiful thing. Our advice is this: Keep some distance between your work life and your home life. Your family and especially your children are important. But make time for one another. It can work. We are blessed with the love of two families—our beautiful blended family and our fire service family.

Firefighters, recognize your extra vulnerability if going through a divorce. Then take steps to keep from utter despair, which can lead to dark, unhealthy, destructive thoughts, especially for men. Men have much higher rates of suicide from divorce than women: "The effect of relationship breakdown can be catastrophic, with most research indicating that men are affected far worse than women. The Samaritans Report found that 'men in mid-life are dependent primarily on female partners for emotional support…and are far more likely to commit suicide after a divorce.'"[11]

Where there's a will there's a way, even a way to get through the difficult aftermath of a painful chapter. It can be done; firefighters are resilient problem solvers. Use your life-saving skills on *yourself*. Seek training ideas with the same passion you would pursue them for the fire service. Here are five to get you started.

Seek help from your fellow firefighters

The high-stress world of firefighting can add to the weight of despair, but the unique bonds of brother/sisterhood can also lighten that load. Do not go through your divorce alone. Tell your crew. They need to know, they deserve to know. Your actions and moods affect them. *Never think for one second that no one cares.* Seek out a trusted officer; they can put you in contact with numerous sources of aid and support (i.e., someone like our particular favorite, SFD Chaplain Joel Ingebritson). You have a second family, and they're not going anywhere. It's wired into their DNA to help, especially to help you. Call upon them and don't feel as if this is an imposition; the bigger imposition will be the heartbreak you inflict if you choose to check out instead. You know they'll blame themselves. As the National Volunteer Fire Council writes, "Firefighter suicide shakes the very core of the fire service, and its impact can be emotionally and mentally debilitating for those who are left behind to grieve."[12]

Warning signs to look for in a fellow firefighter who may be contemplating suicide are: extreme hopelessness/negativity/tiredness, moodiness, buying a weapon (if this is out of the ordinary), giving away possessions, making a will, plans for taking "a long trip," taking unusual risks, substance abuse, or a sudden change in religious beliefs (can include positive or negative extremes). Firefighter, if any of this describes you and you're just too embarrassed to speak about it to a fellow crew member or officer, please do the right thing and seek professional counseling.

Give it time

It may take years to feel normal again after a divorce. It can be as painful as a death; in a way, it is a type of death—the death of a relationship. Don't be in a hurry to end the pain; it simply must be felt, endured, and suffered through. Be patient, it will pass eventually. Take life one day at a time and know—you *will* have joy again.

Never give up on the kids

No matter what the situation, never give up on having a relationship with your kids. The trauma of not seeing them every day can be brutal, but be careful to not make it about what *you* need. What *they* need is to know you'll always be there, in any way, shape, or form they can get. Even if it's only one day a week, hang in there for that one day. If they're hostile, keep reaching out. If the ex is making it difficult, don't quit trying. Never believe that they don't need you anymore. This is an absolute lie. Someday the kids will appreciate your efforts even if they don't right now. I know this from personal experience: my dad kept trying, and I'll always be grateful for that.

He chose to live nearby, he made sure we saw him for dinner every week, and he even took me to lunch as often as possible. And he never forced his second wife on me; I was not compelled to accept her nor did he insist that I see her every time I saw him. If the kids aren't ready for the second marriage, don't force them to be. You'll risk alienating them and that would be your loss. Let them love you. They'll always need you to be there.

Forgiveness is essential

Forgive your ex, no matter how horribly they may have treated you, for your own sake. As Malachy McCourt so aptly said, "Resentment is like taking poison and waiting for the other person to die." Forgiveness does not mean condoning, it means letting go of the *anger* that can turn to hatred which can then lead to rage, even violence. To err is human—remember that. Marriage is tough and a lot of them fail. You might need to forgive yourself as well; maybe the divorce was more your fault. Don't let regret turn to despair.

Learn from your mistakes

There are many reasons for marital failure. Maybe you married the wrong person. Make note of what you felt was wrong about your ex and look for something different in the future. Own up to your failures and determine not to repeat them in the next relationship. Just because one marriage ended doesn't mean a future one will. And you'll have a better chance at success if you choose to change.

Divorce is tough stuff; don't go through it alone. Give it some time, hang in there for the kids, forgive, and apply what you've learned from failure—some of the best success stories are born from the ashes. We know many firefighter couples who have found love after divorce. Sometimes it *is* better the second time around. Life can and does go on. *Divorce is not the end.*

HATED THING...

Way back in the beginning of the book, Anne described my experience while working at SFD's Joint Training Facility. Because my fire department is so large, it's easy to lose track of people through the years and there are many that I'll never even meet. But while at the JTF, I reconnected with a lot of my old colleagues and met many new ones. Since everyone had to come down to train, I saw just about everyone at some point. The steady drumbeat of broken relationships, separations, and divorces was stunning and depressing. Good people with what had seemed like very strong marriages were time and again having them blow up, and many in very bad ways. This truly was the impetus for our book. Anne is not overstating it when she says that we hate divorce. The high number of divorces is a blight on our society and a modern tragedy responsible for so much heartbreak and suffering. I truly hate divorce and all it represents.

I guess the book could just finish here, but that would not be in keeping with thorough and sound counsel. Because the bottom line is that, though we will try hard to keep them from happening, divorce will be a reality for many in the fire service and life will still go on. The kids will still need raising, the lawn will still need mowing, the bills still need paying, and all the rest of life will proceed onward. So as the title suggests, divorce is not the end.

IT AIN'T OVER TILL IT'S OVER...

I know those of you who've been divorced or are currently on that road never planned for it to go that way. Lots of things happened from the moment you said "I do" to the current feeling of "I don't." We've done our best in the earlier chapters to try and keep this from happening.

But it's also really critical that we acknowledge that we're flesh and blood and oh so mortal. We mess up and make mistakes and turn really good things into really big messes. We fail. And with that failure life just doesn't stop. I have witnessed quite a few divorces in my life and here a few things to keep in mind in addition to Anne's advice earlier in the chapter:

DON'T RESORT TO VIOLENCE

I'm sorry to start off with such a heavy thought, but this is so important it could use its own chapter. Emotions are going to be raw, really bad stuff is going to be said and done, finances are going to be ripped apart, and this is just for starters. The inclination to lash out and cause pain is very common and can happen to any of us under these circumstances. Please don't. It will take a bad situation and turn it into something that is almost impossible to salvage. Injury and death are strong possibilities and jail time is likely. Take strong measures to keep yourself from reacting violently and never put yourself in a position where things might get out of control. Reach out to friends to walk through this difficult time and, if necessary, get professional help to deal with emotions. Sometimes the aggrieved spouse will purposely push you to say and do things that are not typical of you. They are in pain too and their behavior may be erratic.

Don't take the bait. Don't get drawn in. Under no circumstances is it acceptable to react with violence, and if you do, the already heartbreaking situation will become unimaginable.

OWN YOUR MISTAKES

There are two sides to your "divorce story," and no amount of disputing that will change it. There may very well be one of you who is more to blame, but rarely does a happy marriage go bad without both contributing to the problems. Own your issues and admit them. This may not fix the marriage, but it gives you a chance to move on and have a happier life. This is especially true if you have kids and will need to figure out sharing time in a way that meets their needs. Being honest about your contributions to the divorce also enables you to have a better chance should you decide to marry someone else.

FEEL THE PAIN, BUT DON'T TURN TO ALCOHOL OR DRUGS

This process is going to be as painful as anything you'll experience in your life. Feelings of anger, depression, confusion, sadness, guilt, and so many more simply mean you are normal. What we hope you'll be wary of is the very real tendency to escape the pain through chemicals, whatever form you prefer. So many horrific decisions and occurrences have happened to very good people simply because they were altered due to alcohol, drugs, or abuse of medications. The temptation is understandable, but please consider other ways of coping. And always ensure your medications are taken under professional medical supervision.

NEVER FORGET THAT YOU HAVE GOOD STUFF AHEAD

As a closing reminder, what is happening now is not always going to be happening. It will have an end. However long and difficult the process, you will have fun again. I don't say this to in any way diminish the terrible nature of divorce. But I know plenty of fine people who endured terrible divorces only to find real happiness in their next marriage. Many of my friends matured as a result of the challenges thrust upon them by losing their first spouse. It's not the most advisable way to grow up, but many have done just that. And it should be a source of encouragement that you can still live a wonderful life and find love and true happiness.

HOPE REMAINS

I will always feel a deep sadness when I hear that folks I know are getting divorced. The stories will vary and details will be different, but the end result is the same. What was once a smiling and happy bride and groom is now an angry, hurt couple going their separate ways. With that sadness will always be a prayer that out of the difficult time, something stronger emerges. Our hope remains that you can find what you lost and rebuild your marriage to the place you'd always dreamed it would be. If that cannot be, remember that life is still going to happen, and it can still be great. Keep rollin'.

NOTES

1. Mark Banschick, "The High Failure Rate of Second and Third Marriages," *Intelligent Divorce* (blog), *Psychology Today*, February 6, 2012, https://www.psychologytoday.com/blog/the-intelligent-divorce/201202/the-high-failure-rate-second-and-third-marriages.

2. Finlay Young, "Why Men Are Killing Themselves," *Newsweek*, February 12, 2015, http://www.newsweek.com/2015/02/20/suicide-men-305913.html.

3. Lecia Bushak, "Married vs. Single: What Science Says Is Better for Your Health," *Medical Daily*, April 2, 2015, http://www.medicaldaily.com/married-vs-single-what-science-says-better-your-health-327878.

4. Elizabeth, "Police, and Other Agencies, Could Take Notes from What the Phoenix Fire Department Has Done," PulseAZ, December 10, 2014, http://www.thepulseaz.com/blog/police-and-other-agencies-could-take-notes-from-what-the-phoenix-fire-department-has-done/.

5. National Institute of Mental Health, "Suicide."

6. LiKamWa McIntosh et al., "Suicide Rates by Occupational Group," 644.

7. Paul Peluso, "Chicago Firefighter Suicide Report Seeks Answers," Firehouse.com News, June 21, 2012, http://www.firehouse.com/news/10732817/chicago-firefighter-suicide-report-seeks-answers.

8. April Kahn, "Suicide and Suicidal Behavior," *Healthline*, reviewed January 27, 2016, https://www.healthline.com/health/suicide-and-suicidal-behavior#Overview1.

9. "Suicide and Divorce," Divorceinfo.com, http://divorceinfo.com/suicide.htm.

10. Ibid.

11. Young, "Why Men Are Killing Themselves."

12. National Volunteer Fire Council, *Suicide in the Fire and Emergency Services* (Greenbelt, MD: National Volunteer Fire Council, 2012), 16, https://www.nvfc.org/wp-content/uploads/2015/09/ff_suicide_report.pdf.

Don't Stop Believin'

I once asked my husband Mike what makes me a good wife. I asked him this not because I was fishing for kudos (OK, maybe I was a little), but because I was also looking for column ideas. He is, after all, my muse—my best source for topics and subjects pertaining to firefighter marriage. Perhaps it was because he just went to a Journey concert but he answered, "You *believe* in me, you've always believed in me, and you've been great at telling me so. I really need that from you."

What does it mean to believe in someone? Believing is seeing—seeing the diamond in the rough that no one else sees, and assuming the best. It is giving someone supportive reassurance even when they fail, *knowing* they can do better next time and telling them so. A supportive spouse gives you another chance for a better outcome, time and time again. I believe a good wife speaks faith into her husband and in so doing sets him up for a favorable result by giving him *hope*. That hope helps him rise up to overcome any obstacles.

In a strong marriage, this belief, of course, is reciprocated. When two people believe in each other, they lift each other up when one of them falls. That's usually all it takes to keep you going, that one helpful person cheering you on. Pity the poor soul who has no one to pick them up, as they are so much more likely to quit in despair.

This quotation says it better than I ever could, which is why we, like so many other couples, included it in our wedding vows: "Love bears all things, *believes* all things, hopes all things, endures all things" (I Cor. 13:7). When you *believe* in your spouse, you see their potential. You tell them all that is good in them, you recognize their hidden strengths and talents, and this is inspiring. Together you are stronger, you build each other up, and in so doing, receive a new and better sense of confidence. As a couple, you can reach greater heights, attain loftier goals, and face adversity without fear, failure without despair.

When you believe in someone, they prosper, and when they prosper, so do you. As they climb they reach back for you and help you up, and vice versa. Higher and higher you go, ascending those mountains and spurring each other on. Successful people always point to having someone in their lives who believed in them. How beautiful it is when that person, that inspiration, is your very own spouse. When the person to whom you bare your soul, reveal all your secrets, and expose all your weaknesses declares you to be absolutely wonderful, you can't help but believe them. When the one who knows you best is truly in your corner, you can't lose. How reassuring this is in the tough arenas of life.

When you believe in someone, you bring out the best in them. Your positive words drown out the cacophony of negativity and criticism because your praise is stronger and nearer and louder. Belief gives forward momentum instead of killing it. *Your* confidence gives *them* confidence to keep moving, striving ever on with the assurance that they surely can because *you* said so!

When you believe in someone, they see themselves through your eyes, a reflection in a loving mirror of the champion they can be. They gain a new sense of courage because you have recognized qualities they themselves never even knew they possessed. And in a good marriage, this mutual discovery goes on and on as you experience new challenges together. For as life changes, so do we. With the support of a spouse we can find hidden strengths that often exceed old ones and on occasion, even surprise ourselves! This is called *growing* old together, not simply aging into decrepitude.

When you believe in someone, they cling desperately to you as a life raft, as they know you are their true source of safety and security. Having learned all their idiosyncrasies and peculiarities and still having declared them to be your perfect match, they know they can trust you for true and pure advice, as you surely have their best interests in mind. You see the good, you see the bad, you see the strength, you see the weakness, *and yet you still believe*—this truly is a source of healing and renewal. Your love, your faith, keeps them from drowning in a pool of fear and self-doubt.

THREE BLESSINGS TO ALWAYS REMEMBER ABOUT FIREFIGHTING

How do I keep believin' in my firefighter? How do I best cope with all that this profession takes from him and from me as his wife? It's easy to focus on the negative, to get stuck in the pitfalls and downfalls of this dangerous, incredible, unique occupation. And sometimes it is necessary to do so, as the difficulties must be identified if they are to be dealt with. But it's just as important to stop once in a while and focus on the positive, to re-examine and remember what is so good about firefighting. It helps to keep your heart thankful, and a thankful heart is a contented one—a believing one. Being a firefighter and being married to a firefighter truly is a blessing and here are three reasons why.

Firefighters love what they do

This is no small thing. Many people simply trudge through their work day with grudging acquiescence, tolerating but not loving what they do. Sadly, lots of folks even hate their jobs and, as Henry David Thoreau describes in *Walden*, "lead lives of quiet desperation," dreaming and wishing for a way out. They're bored, tired, distracted, and disinterested in the drudgery of repetitive, meaningless tasks. But not firefighters: they show up alive and alert, excited to face a day where literally anything can happen. It's exciting, it's challenging, it's never the same. It's physical, athletic, invigorating. Adrenaline flows, hearts race, lungs expand—firefighters know they're alive.

This job offers many avenues; if you're not happy at one station, you can move to another. You can ride a truck, you can ride an engine, you can ride an aid car. You can work in the alarm center or investigate arson. You can promote. You can be a driver. You can be part of special operations teams. The opportunities are nearly endless; the variety offers something for everyone. The schedule is different—sometimes it's a little hard on sleep, yes, but you can work big chunks of time and have big chunks of time off. This is typically more positive than negative for the whole family. When you live with someone who loves what they do, you grow to love it too; enthusiasm is contagious.

Firefighting has meaning

Firefighters save lives for a living. It doesn't get any better than that. They also save homes full of memories and priceless treasures, like photographs and beloved pets. They show up on peoples' worst days and make them better. They restore order to chaos. They return the dead to the living. They get to see positive results from their actions. They are respected and appreciated by the public, and though they don't need it, it's nice to know. At the end of the workday, they feel proud of what they've accomplished and glad for having showed up. Firefighting is a noble calling from God; it is a calling to sacrifice oneself, if need be, to preserve life. This devotion to humanity, this service to the community, is something a spouse can truly respect, and respect always makes marriage stronger. And firefighting is something children can be proud of as well; they believe in you too, firefighter Mom or Dad.

CHALLENGES OF THE FIREFIGHTER MARRIAGE

Firefighters get to work with amazing people

Yes, firefighting is dangerous. There are risks, extreme risks. But firefighters don't work alone; they work with highly trained, effective, capable people who together, side by side, reduce the risks to manageable levels. The job is less dangerous because of the person standing next to you. Deep trust forms when you can put your life in another's hands and know it will be safe there. Firefighters get to experience teamwork on levels most will never comprehend.

At the firehouse, firefighters become more than just fellow employees, they become family. They cook together, they eat together, and they play games to see who has to do the dishes. They pull pranks on each other. When one is injured, the others call, send texts, and even forward encouraging articles like, "Why Redheads Feel More Pain" (firefighter humor) as Mike's crew did when he was on disability. They make each other laugh with their infamous beanery banter. They cheer each other up. And they are truly sad to see each other go when someone is transferred away or retires. It takes a special kind of person to become a firefighter; when you do, you have the pleasure of being surrounded by such people (fig. 24–1). Extraordinary and amazing, caring and strong, fun and enthusiastic—who wouldn't want to work with the best of the best? And what family couldn't get behind coworkers of this caliber for their loved one to enjoy?

Fig. 24–1. Mike's crew—a significant blessing

So be thankful, firefighters and firefighter families: you are part of one of the finest professions in the land. It is work that is enjoyable, it is work that has meaning, and it is work that creates true brother/sisterhood. For these reasons and so many more, firefighting truly is a blessing—a profession you both can believe in.

Rick & Jami Lasky, Lewisville FD, Texas

OUR TWO FAMILIES

The fire service is loaded with incredible people, people who have dedicated their lives to selflessness, to valuing family. It's not just a profession to them, it's a calling. And it doesn't matter whether you do it as a career or volunteer firefighter, it reaches out to people who want to make a difference in the lives of others, to those with a passion to serve.

In doing what they love to do—protecting and serving others—firefighters forget that they now have two families to look after: the "first family," the one at home and the "second family," the one at the firehouse. And the first family, well, that's just what they are: *first*. When you don't have your priorities in order you can end up sliding the second family in front of the first, creating struggles within a relationship. With all of the demands our profession can place on you, such as schooling, training, continuing education, studying for promotions, and even just the time it takes to be good at what you do, it can be easy to forget what's waiting for you patiently at home. We want you to love the job, but we want you to *love your family* more.

So, what can you do to help keep the first family first? Probably the most important thing is to remember that your spouse is your best friend. Your brothers and sisters at the firehouse are family too, the second family, but when it's all said and done and you're ready to hang up your boots, there's only going to be one family waiting for you—the one at home. Sure, you'll see the second family at retirement ceremonies and other department-related functions, but the first family, your spouse and children, are going to be there for you until the end.

Don't ever leave the house without saying I love you. You never know how the day is going to end. Life is way too short. Call your spouse a couple times while you're on shift. And when they call, try to make time for them and not rush them off the phone. It can be busy in the firehouse and it's easier to rush off the phone, but one day you may regret it.

Make your spouse feel welcome at your firehouse. You're already going to miss a holiday or special occasion or two, so do what you can to make them feel comfortable around the second family. If you're a boss, make sure your firefighters' families feel welcome in home number two. Take time to get to know them.

No matter how you paint it, the job is dangerous. With today's media and social media, it doesn't take long for word to travel of a serious fire or worse, one where a firefighter was seriously hurt or, God forbid, lost their life. If it's your department, make every effort to call home and tell your spouse that you are okay. In "the miracle on the Hudson," Captain "Sully" Sullenberger called home to tell his wife that he was okay the moment he made it off the plane, knowing that she would see it on the news within minutes. Whether it's your department or not, your spouse thinks about the dangers of firefighting every day. It's easy to forget who's at home going out of their mind wondering if you're alright.

Both you and your spouse need to know the signs of PTSD. The reality is, it's not normal for a firefighter to see what they see. They can put on a show, but there are signs. Know what they are and what to do if you discover your partner is going through a tough period. Nobody is that tough and everybody has a breaking point. Today's fire service has taken on a much more proactive stance in dealing with and helping those whose may be suffering from PTSD. And if they're the chief, make sure someone is looking out for them. At times, they can become consumed with looking after the men and women in the department and forget to address their own needs.

If your spouse decides to take a day position (Monday through Friday) with the fire department, be ready. We're talking about a major lifestyle change, going from shift work to days, and it can take some work on both ends. In our house, it was a struggle for my wife to go from having her third day of freedom to me being home every night. Thank God, she made it!

Have some fun away from the firehouse. Throughout our marriage we tried to balance our children's chosen sports with making sure we did things we liked to do when we could. Things like going to New York and seeing as many Broadway plays as we could in our short two- to three-day trips. Once the kids were grown we finally did something we'd always talked about: travel to Hollywood and try to get on *The Price is Right*. We never really imagined being called to contestant's row, or winning the whole shebang for that matter, but we did! No matter what, we were going to have fun sightseeing and spending time together.

Some things we both have learned over the years are to love each other with all our hearts, be best friends, be patient through the ups and downs, and yes, those words ring true: "to have and to hold from this day forward, for better or for worse, for richer, for poorer, in sickness and in health, to love and to cherish, from this day forward until death do us part." Life truly is way too short. Enjoy every minute that you can!

Press forward, firefighter couple, with expectancy of a better tomorrow and a brighter future. Believe in each other and in this amazing occupation. This belief gives you hope that your dreams can come true. It keeps you out of the gutters of life and secures you to the path of achievement. With undying, steadfast marital support, you can travel many roads and have incredible adventures as husband and wife.

When someone tells you that you are braver than you believe, smarter than you know, and stronger than you think, you then become willing to dare great things, take bigger chances, and stretch out beyond your comfort zone. Heroism is born of such risk, and who is more heroic than a firefighter? And what is more heroic even still? A loving, lasting marriage that defies the odds. For your sake and for your children's sake, firefighter couples, don't stop believin'—*don't ever stop believin'!*

DIAMOND IN THE (VERY) ROUGH

I chuckle at the notion of being a diamond in the rough. Most folks can easily see the "rough" part of that equation, but Annie truly sees the gem and I love her for it. In fact, she has believed in me when most of the evidence pointed to a less than stellar outcome. All of us need that from our spouse. Not a blind-eyed, pie-in-the-sky vision of stuff that's not real. But the actual potential and qualities that are true and just need edifying. The rest of the world will take its turns knocking you down, pointing out your flaws, whittling away at your self-confidence, and bringing you low. A spouse should be the one person who is determined to see you reach your potential, who never forgets the glorious stuff that is true about you. They believe in you regardless of whether anyone else does or not.

For the firefighter marriage, this is as important as anything else we have written. You've read an entire book about the challenges, struggles, and landmines that lay in the path of a thriving marriage. There will be plenty of things trying to chip away at your happiness and move in to destroy your closeness. Don't give in to these obstacles and let them kill your belief in each other or the dreams you share.

I leave you with a few final bits of encouragement to have each other's back and be your spouse's number-one cheerleader:

As a couple you have every reason to be proud of each other. You as a family are a member of the one of the greatest organizations the world has ever known: the fire service. Yes, it has peculiar challenges. Yes, it presents hazards, both physical and emotional, to being a happy couple. But friends, the fire service is simply awesome.

It is revered and sacred. When members of your community are having the single worst day of their lives, they call upon their firefighters. Your firefighter. For it all to work, the firefighter needs to be in top form, and one of the best ways for that to happen is with a loving, supportive spouse. Remember the power you have as a team and believe in the impact you can have together.

Believe in each other, because there will always be those who want to tear you down. It will surprise you where this will come from and who will be working against you. Many people in your world will be personally unhappy and more than ready to see you share in that condition. Misery thrives on shared despair and, whether knowingly or not, others will look to bring you down. That's one of the beautiful aspects of two becoming one. It's why a couple united and moving in the same direction is so powerful. I love the imagery of a husband and wife standing side by side while the world rages all around them (as pictured on the cover of this book). Holding each other up, encouraging each other to never quit. That's the good stuff, my friends. That's the heart and soul of believing in each other.

Find what it is your spouse truly dreams of and help them to live that dream. You are holding in your hands my beloved bride's dream. She has always been an amazing writer and quietly dreamed of someday publishing a book. It is always recommended that you "write what you know," and Anne certainly knows what being a great spouse is all about. What she did not know was if being published was even possible. She needed someone to believe in her, and I did and do. There were others who were helpful on this journey, but it truly took me, at her side, to believe that she could succeed and I told her so. Repeatedly. Her dreams are a passion of mine every bit as strong as the ones I have for myself. I have watched Anne do the exact same thing for me and our boys for the 30-plus years I have known her. It is an incredible blessing to see the dream realized for the most caring and unselfish person I have ever known.

LOOKS LIKE WE MADE IT

Every New Year's Eve, Anne and I do the exact same thing. We hold each other, share a kiss, and say, "We've made it another year." Despite all the challenges the world, firefighting, and life itself have thrown at us, we've made it. Finances are always difficult, health varies, and the family has its ups and downs. Our work is at times amazing and at others a drag. In short, our lives are just like yours. But we have held each other tight, kept each other going and believed in each other's dreams. We've never given in to the culture's mad dash to seek self first at all costs. What we dreamed of having, when we knelt before God and pledged our lives to each other, remains our goal. The challenges of firefighter marriage are real. Our desire to love each other and enjoy life remains our strongest weapon in meeting those challenges. We are determined to win.

Chapter 24 Don't Stop Believin'

Our prayer is that you will find encouragement and answers to some of your own challenges in this book and that you'll never stop believing in your firefighter marriage. Believe in your own dream and don't ever stop.

Fig. 24–2. Future heroes never stop believin'. Courtesy of Tanya Miller.

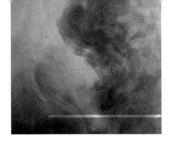

Conclusion

As a concluding note to what we hope has been an enjoyable, meaningful, and beneficial book, we hope that you'll reach high in your marriage. Don't settle for mere cohabitation or being "friendly roommates."

Strive for the passionate intimacy you've heard can happen. It can.

Commit to selfless service of your spouse and meeting their deepest needs. You will.

Build a lifetime of trust and sharing that will be a more significant accomplishment than anything else you pursue in the fire department, or in life. It's worth it.

Be a soul mate, not just a "sole mate." This is truly a dream worth dreaming...and Anne will tell you why...

— *Mike*

SOLE MATE VS. *SOUL* MATE

A mate. We all long for a mate, a companion, a match that turns our single into a pair. Someone to hang with, talk to, and hopefully, share the deepest of physical intimacy. Spend holidays with. Someone who keeps us from being alone. A trusted friend, a committed relationship. A mate.

But there isn't just one kind of mate; there are, in fact, two. They sound the same, but in reality, in meaning, they are infinitely different. A few questions you should ask yourself as you read further are these: which kind of mate do you prefer to have, and which kind are you willing to be, a *sole* mate or a *soul* mate?

A sole mate. Sole, singular, exclusive, the only one. I will have sex with just you and no other. Monogamous and committed to be so for now, and possibly longer if things go well. I will share with you and only you my physical presence for as long as the relationship lasts.

A *soul* mate. The soul—the spiritual part of a person, their morals and emotional warmth. Their vital force or essence of heart, substance, individuality, and disposition. This is what I will share with you: my deepest thoughts and nobler qualities of affection, honor and duty, poetry, reverence, and beauty. I will long for your company as well as your touch. We will have a connection beyond that which is just physical. And I will share it with you, all the days of my life.

A sole mate. I will live with you in comfort and safety. We'll build a house and have nice things. We will establish boundaries and ownership and work out the business of life. We'll function as coworkers, each will do their share. A partnership: I'll do my part if you do yours. But I will always hold something back, in case the relationship fails, in case you don't hold up your end of the bargain. Safe, not too risky, comfortable. And it's nice not to be alone. An exit plan in the wings, just in case, as nobody's perfect.

A *soul* mate. Our love is a bright star that shines in the darkness to guide us home, which is never a place but simply wherever *you* are. And in our home, I will share with you all the blessings that come our way and endure whatever misery we encounter firmly by your side. Your good is my good, your pain is my pain. We may be poor, we may be rich, and we may be everything in between; it does not matter to me. All that I have is yours. Your happiness is more important to me than my own. My happiness is to be forever with you—my home.

A sole mate. Our communication is necessary to cohabit. Perfunctory, brief, friendly. Our darker thoughts, we keep to ourselves. Our deepest dreams are just that—deep, unknowable, best kept private. Our struggles are our own. Your work is your business; my work is my business. We have many friends with whom we can share; no need to dump all your problems on me. We keep it light, we keep it easy, we keep it fun. Anything more and it's no longer comfortable.

A *soul* mate. With one glance, I know what you're thinking. I can't wait to hear what you have to say—about everything. Each word, every gesture conveys meaning that only I can read. We finish each other's sentences. We know each other's stories, inside jokes, a secret language, a catalogue of shared memories available to draw upon. Our lives so intertwined I no longer know if that memory is from your childhood or mine. I was not there, but I experienced it with you just the same. That's how well I know you.

A sole mate. We avoid conflict like the plague that it is. No harm, no foul. We agree to disagree—peace at all costs. Not a lot to argue about, as boundaries are firmly placed. We're good friends—cordial, respectful, but guarded. After all, you can't surrender all that you are to another person—that would be foolish. You have to have pride and a strong sense of self. You stay in your space and I'll stay in mine; we'll come together for sex. If problems arise, one of us will simply have to walk.

Conclusion

A *soul* mate. So in sync are we that even the smallest of infractions feels monumental. We're in agony if all is not right, if we're not heard, if we're misunderstood. I cannot function if distance forms between us. The world tilts off its axis if anything is amiss, I will not rest until we are again as one. A night apart would be unthinkable; I cannot sleep until all is forgiven, your hand resting in mine. To you I will bare my soul, my body, my will—and submit wholly, lying exposed and vulnerable at your feet, knowing I'm completely safe there.

A sole mate. We have warmth, we have company, we happily coexist. We even have fun. Ours is a relationship that works. We are bound together by a piece of paper, nothing eternal. We are animals that feel compelled to mate, so we do. And we conform to society's rules as needed, especially when children are involved. But moving on may be necessary if too much of me is being lost to you, or if my needs are not being met. Someone younger and better may come along. Survival of the fittest. I will never risk too much; my best interests must always come first. You are my sole mate because you are of benefit to me, and there is no one else. If that ceases to be the case, I will end this, or you will—that is the law of the jungle.

A *soul* mate. Ours is a lifelong romance, a passionate, never-ending love affair. You are my ideal, my fantasy, my dream come true. Our decades-long marriage has taken on a life of its own—a spiritual entity greater than just the two of us individually or together.

> *Our marriage shines around us and between us with an otherly light, a sacred habitation for our shambolic humanity.* It is soul stuff made visible.
>
> —Andrew Klavan, The Great Good Thing

Our long love reveals a realm beyond the ordinary—the miraculous, the eternal, the divine. And the fact that such a marriage is possible leads us to belief in a God who designed us so perfectly for each other and then with His guiding hand, joined us as one flesh. You are my destiny. My one true love. *My soul mate.*

Which will you be? Will you settle for *sole*, an exclusive mate only, one that works but has no real spiritual depth? It's all very nice, it's all very easy, but not very good. Or will you aspire to that next-level, passionate, intimate connection that brings fire and electricity and meaning to your marriage? It is the greatest risk of all, to bare your soul to another—to lay it all down on the altar of self-sacrifice. Sole or soul, firefighter couples—*you decide.*

—Anne